O. W. Smith

How to Restore Health; a Manual for the Family, Traveler and

Student

O. W. Smith

How to Restore Health; a Manual for the Family, Traveler and Student

ISBN/EAN: 9783337210625

Printed in Europe, USA, Canada, Australia, Japan

Cover: Foto ©berggeist007 / pixelio.de

More available books at **www.hansebooks.com**

HOW TO RESTORE HEALTH;

A MANUAL

FOR THE

Family, Traveler and Student,

BY DR. O. W. SMITH.

———◆◆———

UNION SPRINGS:
J. B. HOFF, ADVERTISER PRINTING OFFICE,
1877.

INTRODUCTION.

HISTORY.

No apology is needed for this work. It is not designed to take the place of the intelligent Homeopathic Physician, but to give a safe and efficient mode of treating the more common ailments of life in place of the unwise and often hurtful measures so generally in vogue in domestic practice. Nothing need be said to the patrons of Homeopathy to convince them of its superiority. To the inexperienced, doubting, or prejudiced I would simply say, *try it*. Give the method a fair test before condemning it. A short history of Homeopathy, with a brief explanation of its doctrines may not be out of place.

Its founder, Samuel Hahnemann, in 1790, while translating Cullen's "Materia Medica" from the Latin into the German language, was struck with the contradictory properties ascribed to Peruvian Bark, and the different explanations of its action in intermittent fever. He tried the effects of the drug upon himself, and found it to produce symptoms *similar* to those of intermittent fever. This fact suggested to him the principle *Similia similibus curantur*, (Likes by likes are cured). He thoroughly tested this principle upon himself and others with a variety of drugs; and having obtained similar results in every instance, he announced the principle and began to apply it to the treatment of the sick.

Like every reform this passed through a fiery trial. Hahnemann's suggestions were treated with scorn, and for fifteen years he was the object of ceaseless attacks from those whose interests were opposed to the changes he sought to introduce into medical practice. This persecution finally drove him from Leipsic, where, as well as at Berlin, stand to-day statues in his honor. Every Physician who received his doctrines was insulted, denounced and ostracized by his professional brethren. No darker page exists in the history of ecclesiastic persecution than is found in the history of the intolerance and bigotry of Allopathy toward Homeopathy. Yet, in spite of the most intense and bitter opposition Homeopathy has steadily and permanently advanced into the confidence and affection of the people. In the United States alone it has grown during the past fifty years, from one Physician to six thousand, the majority of whom are graduates of Allopathic schools. It has ten regularly chartered medical colleges, which in their thoroughness of teaching compare favorably with those of the other system. It has several large hospitals and asylums, and many dispensaries. It has eight well sustained journals. It can number its patrons by the hundred thousand from among the refined, intelligent, and thoughtful people of this country. In England, France, Germany, Russia, Holland, Austria, Spain and Brazil it is making steady progress among the intelligent classes.

DOCTRINE.

The essential doctrine of Homeopathy is this : *The best method of curing a natural disease is to give a remedy, which, if given to a person in health, in large or poisonous doses, would produce symptoms similar to those of the natural disease.* Abundant proof of the truth of this is found in Allopathy itself. Ipecac is given in large doses to produce nausea and vomiting, yet Allopathists give it in small doses to *cure* nausea and vomiting. Drs. Trouseau and Pidoux, two eminent and modern French Physicians of the Allopathic school, state that "When Hahnemann promulgated his therepeutic formula, *Similia similibus curantur*, he supported his assertions by citations from the practice of the most illustrious Physicians. There is every proof that local inflammations are frequently cured by the direct application of irritants which cause similar inflammation, the artificial irritation substituting itself for the primitive one." Dr. Symonds, an Allopathic Physician of high rank, in the "Cyclopedia of Practical Medicine" says: "Upon this ground we were disposed to suggest the use of strychnia in tetanus, not that we have become followers of Hahnemann, but that it is a simple and undeniable fact that disorders are occasionally removed by remedies which have the power of producing similar affections." Dr. George B. Wood, in his "Therepeutics" says, "The same medicine may produce opposite effects in health and disease. Thus cayenne pepper, which produces in the healthy fauces redness and burning pain, acts as a sedative in the sore throat of scarlet fever. A concentrated solution of acetate of lead acts as an irritant, while the same solution very

much diluted will act as a sedative." Carbolic
acid will often produce vomiting, even when absorbed from a wound, and flatulence is a nearly
constant symptom with those who have taken or
proved the drug, yet Dr. H. C. Wood recommends
it for vomiting and flatulence. The nitrite of
amyl will produce sudden dilitation of the arteries,
yet Dr. Ringer found it of the utmost service
where this condition existed, the *flushings* of
women at a certain age. Dr. Ringer also speaks of
the use of tarter emetic in bronchitis, and says
that the nausea and vomiting are the first symptoms to yield to its use ; yet the power of Tarter
emetic to produce nausea and vomiting is well
known. Moreover, in the poisoning of animals
with this drug, Majendee, Lepelletier and Molin,
found the lungs always more or less affected, from
simple congestion to hepatization, yet it is the
standard remedy with Allopathists for congestion
and inflammation of the lungs. Arsenic, in poisonous quantities produces inflammation and ulcera-
tion of the stomach and bowels, yet Dr. Hunt says
that arsenic benefited his patients who suffered
from a similar trouble. Dr. Thorowgood extols
arsenic in " Irritative Dyspepsia ;" and Dr. Ringer
recommends it in the morning vomiting of drunkards, and in chronic ulcer of the stomach. In
chronic poisoning by arsenic a variety of eruptions
is produced, yet it is more frequently used by
Allopathists for skin diseases, than any other
remedy. Similar facts can be related of many
other drugs. In vaccination or inoculation we
have a most convincing illustration of the
Homeopathic law. In the light of all these
facts who will say that Homeopathy is a delusion
and a humbug?

It is impossible with our present means of

observation to ascertain *how* "likes by likes are cured." We cannot look into the human body and see the vital and chemical actions going on; but in the operations of the forces of nature *outside* of the body, we may get some light upon this point. Throw two stones of equal size and weight into the water a little distance apart and you will see the waves from one completely *neutralize* those from the other. Suspend a steel bar between two magnets of equal attractive force, and you will find those forces to *neutralize* each other, and the bar of steel at an equilibrium. Two rays of light of the same wave length passing in the same direction *neutralize* each other, and we have darkness as the result. Sound waves of the same length also *neutralize* each other. I believe it is upon this principle of *neutralization* of similar forces that diseases are cured by remedies. The morbid or disease producing force is met and neutralized by a similar acting remedial force.

With many the *small dose* of Homeopathy is a great stumbling block. It is difficult for such to distinguish between *quantity* and *quality*; and a standard argument with most Allopathic Physicians against Homeopathy is, that some child has eaten a box of pills and received no harm; and *of course* a medicine that does not make a well person sick cannot make a sick person well. I think a little reflection will convince any candid mind that this argument is very flat and weak. Many of the most fatal diseases, such as diphtheria, scarlet-fever, typhus-fever, erysipelas, puerperal fever, small pox, yellow fever and cholera, are due to distinct and specific poisons. These poisons may be burned up, sealed and carried in a letter, transported by ship to distant countries, conveyed in the clothing, and carried on the winds;

showing thus that they are subject to physical laws, yet so minute and intangible are these powerful poisons that no microscope has been able to make them visible. The quantity of virus sufficient to produce any one of the above named diseases is so small that it is absolutely beyond our power to detect it. Now, is it unreasonable to claim that it will require no more of the antidote to cure a disease than it requires of virus to produce it? Professor Taylor tells us that the quantity of lead required to produce the symptoms of lead poisoning is *inappreciable*. The emanations of arsenical wall paper may so poison the atmosphere of a room as to produce very serious symptoms in the persons breathing it, yet the most delicate chemical test would detect no arsenic in the air. Thus we see that a very profound and serious change may be wrought in the human body by the *constant* action of an inappreciable quantity of a drug. The dose of Homeopathy is a strong argument in its favor. If a mistake is made in giving Homeopathic remedies, no permanent harm will be done; but with the large and poisonous doses of Allopathy a mistake has more than once proved fatal. In disease the organ or part affected is far more sensitive to the action of a drug that has a special affinity for that organ. When the eye is in a healthy state the bright and beautiful sunlight affords pleasure; but let the eye be inflamed, and a ray of sunlight will cause the most intense pain. How sensitive a person may be to the slightest touch or breath of air when afflicted with rheumatism. How painful the sound of music to the congested brain; and how a diseased stomach may suffer from the blandest food. These illustrate the increased sensitiveness of diseased organs, and this susceptibility exists in relation to drugs. Experience has shown that a quantity of medicine sufficient to cure a disease might have no visible effect in health. Homeopathy does not debar a Physician from any measures that may aid him in curing his patient. Light, air, water, heat, cold, electricity, exercise, diet, may all be brought into service. Nor is he debarred from using drugs for their mechanical or chemical effects. He may give an emetic or purgative to expel some substance from the stomach or bowels; or a styptic to arrest bleeding; or a vermifuge to expel worms. He may administer antidotes to poisons, and in such quantities as the laws of chemical affinity make necessary. He may give drugs to effect some chemical changes in the fluids or solids of the body, or to supply some lacking element. All this has, however, no bearing upon the distinct doctrines of Homeopathy or Allopathy.

Homeopathy is *pleasanter*, *safer*, and *more successful* than Allopathy. Who, that has ever witnessed the heart-sickening struggles of the *little ones* against the horrible mixtures of Allopathy, will not grant it to be *pleasanter?* Mothers all over the land bless the system that saves their loved ones and themselves so much needless suffering. That it is the *safer* method must be self-evident. All drugs are essentially poisons. They are not, of themselves *restorative*. Given in health they always do more or less mischief. Both classes · of physicians avail themselves of the poisonous action of drugs. The Allopath gives large doses in order to get up a *genuine drug disease* which he hopes may be *substituted* for the natural disease. The Homeopath finds a very minute dose sufficient to get up an influence that will *neutralize* the natural disease; thus restoring equilibrium, harmony and health. Certainly

that method must be safest that succeeds in
curing with the least amount of drugging. Allopathy would stand aghast were it faced with
its victims. How many have slept the sleep that
knows no waking from over-dosing with *Morphine*. How many have been brought into a de-
cline from which they never rallied, from profuse
bleedings. How many broken constitutions from
Mercury. How much of *drunkenness* and *opium* or
morphine eating may be traced to *tonics* and *anodynes*.
Homeopaths do not believe in making the whole
body sick to relieve a single organ. They have
learned by testing upon the healthy body, the
organ or part that each drug has an affinity
for ; and by this knowledge avoid disturbing those
organs that are not diseased.

———

STATISTICS.

That Homeopathy is more successful than
Allopathy you may have the evidence of every
Physician who has practiced both systems.
Moreover, it can be shown by reliable statistical
proof. The average duration of disease in the
Allopathic hospitals of Paris, Berlin, Gottingen
and Suttgart is 29 days. The average duration in
the Homeopathic hospitals of Vienna, Munich and
Leipsic is 21 days. In the years 1857, 1858 and
1859 the medical department of the State Pris-
on at Jackson, Mich., was under Allopathic
control. During 1860, 1861 and 1862 it was under Homeopathic control. Again in 1870 and
1871 it was under Allopathic control ; and in
1873 and 1874 under Homeopathic.

Below is a tabulated comparison of the results

Under Allopath control. 1857, 1858, 1859.	Average number convicts pr year. 435.	Total deaths 39.	Total days labor lost, 23,000.	Total cost of hospital supplies $1,678.00.
Under Homeopath control. 1860, 1861, 1862.	545	20	10,000	$500.00
Allopath 1870, 1871.			24,000	$1,800.00
Homeopath 1873, 1874.			11.000	$900.00

The principal cities of the United States have each a Board of Health which keeps a record of all the deaths occurring within its jurisdiction. Their records contain age, nativity, cause of death, and name of attending Physician. The records of New York, Boston, Philadelphia, Newark and Brooklyn, extending over four years, have been analyzed, and the results are given in the tables below. In these results are excluded all deaths in hospitals, also those from still-birth, accidents and violence. The results of private practice alone are compared.

Let us first take the mortality of

NEW YORK CITY.

Year.	No. of Physicians.	No. of Deaths.	Average deaths to each Physician.
	ALLOPATHIC.		
1870....................	944	14,869	15.75
1871....................	984	15,526	15.78
Total...............	1,928	30,395	15.76
	HOMEOPATHIC.		
1870....................	143	1,287	9.00
1871....................	156	1,243	7.97
Total...............	299	2,530	8.46

Here we find the startling revelation that in New York city the average Allopath loses nearly 16 patients annually, while the average Homeopath loses less than 9! This, too, when both are practicing side by side in the same locality, subject to exactly the same epidemic, malarial, and climatic influences.

Let us now take up the mortality record of

BOSTON.

Year.	No. of Physicians.	No. of Deaths.	Average deaths to each Physician.
	ALLOPATHIC.		
1870	218	3,872	17.76
1871	233	3.369	14.46
1872	233	4.575	19.63
Total	684	11.816	17.27
	HOMEOPATHIC.		
1870	40	402	10.05
1871 ....:	44	363	8.25
1872	54	446	8.26
Total	138	1,211	8.77

Here again we are confronted with the same astounding result; the Allopathic losses by death are to the Homeopathic more than 17 to 9!

PHILADELPHIA.

Year.	No. of Physicians.	No. of Deaths.	Average deaths to each Physician.
	ALLOPATHIC.		
1872	655	12,468	19.03
	HOMEOPATHIC.		
1872	168	2,162	12.87

This fatal year shows the same wonderful disparity.

BROOKLYN.

Year.	No. of Physicians.	No. of Deaths.	Average deaths to each Physician.
	ALLOPATHIC.		
1872	317	7,636	24.08
1873	333	7,181	21.56
Total	650	14,817	22.79
	HOMEOPATHIC.		
1872	84	976	11.62
1873	92	916	9.95
Total	176	1,892	10.75

NEWARK, N. J.

Year.	No. of Physicians.	No. of Deaths.	Average deaths to each Physician.
	ALLOPATHIC.		
1872	77	2,121	27.54
1873	77	1,185	15.39
Total	154	3,306	21.46
	HOMEOPATHIC.		
1872	13	168	12.92
1873	16	153	9.56
Total	29	321	11.07

And due allowance being made for the respec-
tive numbers of physicians, we find the ratio of
deaths under the two systems to be as follows:

DISEASES.	DEATHS.	
	Homeopathic.	Allopathic.
Bronchitis...........................	48	100
Cerebro-Spinal Meningitis..............	44	100
Cholera Infantum.......................	64	100
Croup.........................	37	100
Diarrhœa	35	100
Diptheria	63	100
Dysentery..........	39	100
Erysipelas	33	100
Inflammation of Brain	69	100
" " Bowels...............	33	100
" " Lungs	39	100
Scarlet Fever.........................	69	100
Small Pox.............................	61	100
Typhoid Fever........	88	100

It is worthy of special note, that in regard to small pox, we know not only the number of deaths, but also the exact number of the cases treated by both schools; for the Health Board requires a report of every case, whether fatal or not. In this disease, therefore, we have the precise ratio between cases and deaths; and the result confirms the accuracy of the general statistics already given.

In whatever way, or under whatever circumstances, a practical comparison is made between the schools of medicine, a similiar result is obtained. We can but refer to the fact of the acknowledged success of Homeopathy in veterinary practice; and the late epidemic horse-distemper, or "epizootic," added fresh testimony to its superiority.

How can we account for this? Is there any fallacy in it? Do these figures tell the whole truth? In reply, this question suggests itself: Do the Homeopaths treat as many patients, proportionately, as the Allopaths? What is the ratio between the number of patients treated at the two schools, and the number of deaths given in these tables?

This query, which at first sight seems vital, proves, upon examination, to be of little or no practical importance. We could not honestly and fairly compare the mortality occurring in the practice of any two physicians as a test of their relative success, unless we really knew how many patients each had treated during the year; but when we compare the two schools of practitioners in a mass, thus including hundreds, and even thousands, of every age, and grade, and degree of ability, we are safe in assuming that the average Homeopath on one side treats as many patients per annum as the average Allopath on the other; and that this is a fair assumption will be readily believed by any one who will compare the apparent business success and thrift of the two classes of physicians. Consequently we believe and maintain that these tables of mortality, as they stand, are a fair exponent of the relative merits of the two medical systems.

An average Allopathic mortality almost twice as great as the Homeopathic! How can this additional evidence be explained away? We have heard it urged that Homeopathic physicians, as a rule, are called upon more frequently than their Allopathic brethren, to prescribe for trivial ailments, and that thus their time is taken up with cases that rarely become serious and endanger life. The unsoundness of this charge is apparent, when we reflect that the physician of the family attends to all the cases of illness that occur in that family; and the reason that the Homeopath seems to have more trivial cases to attend, can be found in the fact that he does not often convert a trivial case into a serious one. Those physicians who have practiced for years on both systems universally attest that, though they now, as Homeopaths, visit daily more patients than they did as Allopaths, they are much less frequently called out at night to look after the disturbing effects of medicine given during the day. Of course, they have fewer serious cases, because they do not make them serious; and consequently they have fewer deaths.

The accuracy of the above tables has never been denied. Indeed, it were useless so to do, for the records from which the figures were drawn are on file in the public offices of the cities, and their truth can be substantiated at any time.

READ CAREFULLY THE GENERAL DIRECTIONS.

Each name will be mentioned by both the common and technical names; the first in Roman Letters, the last in *italics*. These will be placed in the index, so they may be found readily. A short description of each disease will be given. Under TREATMENT will be given, first, such means as are not strictly medicinal; following this, the *remedies* with the symptoms that indicate them. *Do not forget that the best remedy is the one that is most similar to the condition of the patient*, even if it should not be mentioned in connection with the particular disease. In trying to select a remedy read the entire section and any others that may be referred to. A *Materia Medica* will be found in the body of the work which should be studied.

HOW TO CARE FOR AND GIVE THE REMEDIES.

The medicine case should be kept in a cool, dry place. It should not be the plaything of children. The corks should not be changed; nor should a cork be used that has been in contact with any other substance or remedy. A vial should not be used that has ever contained any other drug. The medicines should not stand near any strong smelling substances, particularly camphor. Keep the vials well corked and handle with clean hands. For internal use dissolve *six* pills in six teaspoons-

ful of pure water. When the remedy is in powder
dissolve about as much as will lie on the point of
a knife blade. Perfectly clean glasses and spoons
should be used, and one spoon for each remedy.
Glasses containing solution should be covered and
not exposed to strong odors. Medicines in solution should not be kept in the sick room. DOSE:
One teaspoonful for an adult; 1-2 for a child; 1-4
for an infant. Repeat the dose every one-fourth
or one-half hour until the symptoms abate if they
are very violent and dangerous, especially in *croup*,
neuralgia, colic, convulsions &c. In dysentery,
diarrhea &c., a dose may be given after each movement of the bowels. Generally, the dose should
be repeated every three or four hours in acute diseases; and two or three times a day in chronic.
Often two remedies are given in alternation; but
I believe the better way is to give the remedy
which the more prominent and dangerous symptoms call for, and continue it until those symptoms have abated; then give another remedy if it
seems indicated. Having made a careful selection,
do not be in too great haste to change a remedy;
if, however, the patient grows worse, or if there is
no improvement within 12 or 24 hours, select some
other remedy. Any remedy can be obtained by
mail by enclosing 25 cents in a letter to Dr. O. W.
Smith, Union Springs, Cayuga Co., N. Y. For
medical advice and remedies, state *all* the symptoms,
age, sex, temperament, color of hair, eyes, figure,
previous history &c., *enclosing* two dollars.

GENERAL DIRECTIONS.

Put the patient in the most favorable condition
for getting well. Give him the most pleasant

room in the house if possible, into which plenty of air and sunshine may be admitted when desirable. Ventilation of the sick room should be constant, or nearly so. When possible raise the lower part of one window, and lower the upper part of another; or the lower sash may be raised a few inches and a strip of board nicely fitted beneath it ; this will leave a space between the two sash through which a current will enter and pass upward. All this must be regulated by the force and direction of the wind. In a serious case a thermometer should be used and the temperature kept even. The bed should not stand against the wall or ceiling. The bedding should be changed and aired frequently, and if possible hung in the sunshine. Fresh bedding should be thoroughly dry and warm so that the patient may get no chill from it.

The patient should be bathed often, every two or three days at least. This should be done before the bedding is changed. Before bathing, the temperature of the room should be brought up and allowed to go down afterward. Warm water with *unperfumed* soap should be used. · The bathing may be done under the covering if thought best, wiping the patient dry afterward. Slamming of doors, stamping, romping and shouting of children are particularly unpleasant to the sick, and should be prohibited. The untimely and ill-advised visits of neighbors and friends often do much harm, and should not be permitted. All odors, tobacco smoke, camphor, and smell of cooking food should be excluded from the room. Let in the sunshine and air the room whenever it can be done without injury to the patient. There should be no whispering nor secret maneuvers in the hearing or sight of the patient; they are very annoying to a sick person.

Never arouse a patient from a healthy and natural sleep. Food for the sick should be prepared with the utmost care and cleanliness. It should not be urged upon the patient if his stomach revolts against it. All smoked meats and highly seasoned food should be abstained from generally. Should there be a *craving* for any particular food, it may be allowed, very moderately at first.

A nurse should be gentle, kind, cheerful, firm, self-possessed and *clean.* His *breath* and *clothing* should be free from the odor of tobacco. All teas, snuffs, castor oil, soothing syrups, coffee, tobacco, spirits, patent medicine &c., should be abstained from while taking Homeopathic remedies.

Section 1. Apoplexy.

Apoplexy is a hemorrhage (bleeding) within the brain. This bleeding may be sudden and profuse from the rupture of a vein or artery, it may be gradual—a sort of an oozing from the very minute blood-vessels. From various causes, among which are intemperance, the immoderate use of tobacco, and sexual excesses, the coats of the blood-vessels loose their toughness; a rupture may then occur from violent coughing, laughing, vomiting, or straining. When caused by the bursting of any considerable vessel, the person attacked falls suddenly, if he be standing; the face at first is pale; the breathing is generally labored and snoring, with a puffing or blowing when expiring; in such cases the face becomes purple. Voluntary motion ceases; the limbs hang motionless. Should the person survive this attack, symptoms of inflammation of the brain manifest themselves. In cases from simple *oozing*, no such sudden attack occurs. Partial paralysis is the result.

TREATMENT—Place the patient in a comfortable position, where the air can freely circulate. Loose the clothing about the neck and chest.

Belladonna—Loss of consciousness. Red face. Pupils dilated. Spasms in the face. Difficult swallowing. Passes urine unconsciously. Reaches with the hands to the privates. Paralysis of the limbs.

Cocculus—Face red and hot. Dilated pupils. Eyes closed with balls rolling about under lids. Breathing quiet. Paralysis.

Hyoscyamus—Sudden falling with a shriek. Involuntary discharge of stool and urine.

Opium—Breathing slow and snoring. Unconscious, with open eyes. Pupils contracted. Slow pulse. Hot sweat on head. Foam before the mouth.

2. Brain Fever. (*Encephalitis.*)

The causes of inflammation of the brain are various. Blows or injuries about the head, Poison in the blood, result of apoplexy. Anything that produces stagnation of the blood in the brain may cause it. Symptoms at first are headache; sensitive to light, noise or touch; flickering before the eyes; noises in the ears; restlessness; sleeplessness; jerking of limbs; grinding the teeth; convulsions; delirium. Lastly: insensible to noise, light, or pressure; pupils dilated; breathing slow and snoring; stupor.

TREATMENT—Bathe the feet and limbs in water hot as can be borne. Darken the room. Prevent all noise.

Aconite—In the beginning when there is a hot, dry skin, quick pulse, great restlessness, anxiety and fear, fear of death, predicts the day he will die, dizzy or faint when rising up.

Belladonna—Beating in head; eyes red and sparkling; pupils dilated; face red, hot and bloated; beating and throbbing in arteries of neck; delirium furious, wild; wants to bite and strike those near him; starts and jumps during sleep. Belladonna is suitable in most cases.

Bryonia—Patient feels as if brain would burst through the skull; delirium worse at night, talks of business of the day; thirst for drink much at a time; lies quiet, as the least motion makes him worse; faint and nausea when sitting up, obstinate, passionate, irritable mood; great pain in joints and limbs prevents motion.

Opium—Unconsciousness, eyes half closed; pale face; deep stupor; pupils very small; breathing slow, snorting, stertorous; delirious talking; eyes wide open, face red, puffy; dull, stupid, as if drunk ; urine suppressed, costive, stools in round, hard black balls.

Other remedies, *Hyoscyamus, Stramonium, Rhus. tox.*

---

3. Epilepsy. *(Falling Sickness.)*

Epilepsy is a kind of fits that occur repeatedly, often periodically; and are characterized by loss of sensibility and consciousness, and attended by spasms. This disease is frequently inherited by children from parents; and there is good reason to believe that the immoderate use of *drink* and *tobacco* in the parent often bestows this curse on the offspring. The French physicians have put down as characteristic of Epilepsy: 1st. Shriek; falling to the ground; deadly paleness; spasms, lasting from one-fourth of a minute to a minute. 2nd. Redness of the face; convulsions; insensibility, one and a half to two minutes. 3rd. Gradual subsidence of convulsions, three to eight minutes.

TREATMENT—During the fit place the patient where he can get plenty of air; loosen clothing about neck and chest; prevent him from doing injury to himself.

Agaricus—Twitching of eye-lids and eye-balls; trembling of legs and hands. Worse before a thunder storm. Burning and redness of fingers and toes as if they had been frozen.

Aethusa—(See under convulsions.)

Arsenicum—Before the fit, a feeling of warm air streaming up the spine into the head. During the intervals, pressive pain in back of head, burning in the spine, sweet taste in the morning, diarrhea with burning at anus, frequent cramps in calves of legs.

Belladonna—Convulsions commence in the arm, bends the body stiffly forward and then backward, clenches the teeth together. Before the fit, congestion of the head. During interval; dizziness; red face; pupils enlarged; rush of blood to the head; ringing in the ears ; jerking in sleep.

Cocculus—Pupils dilated, eyes closed, eye-balls roll around beneath the lids. For women who are very nervous with suppressed or painful menses.

Glonoine—Blood seems to rush from chest to head. During spasms spreads his fingers and toes apart.

Hyoscyamus—Before the attack : dizziness; sparks before the eyes; sensation of gnawing and hunger in the pit of the stomach. *During the fit :* face purple; eyes projected; shrieks; grates his teeth; discharges urine and stool. *During the intervals :* tearing and beating in right eye, which weeps and seems projected. The attempt to swallow fluid renews the attack.

Ignatia—In children, after punishment, fear or fright. With frothing at the mouth and kicking with the legs. Returns daily at the same hour.

Ipecac—Shrieks. Bends the body back. Face pale and puffed. Derangements of stomach. Body stretched out, stiff, followed by spasmodic jerkings of the arms.

Nux Vom—Painful spot in the abdomen, back of navel; pressure on this spot renews the attack,

Opium—Fits occur in the night; with mental derangements. Throws or stretches limbs at right angles with the body ; or, bends the body stiffly backward and rolls sideways. Spasm begins with loud scream; then foam at the mouth; trembling of the limbs; suffocation; eyes upturned. After attack : deep sleep; face deep red and hot.

Stramonium—Paralysis of one side, convulsions of the other. Thrusts the head continually to the right. Continual motion in a circle with the left arm. Rotates the arms over the head. Convulsions excited by bright, dazzling objects, water or touch. Low spirited. Fear of death. Desire to be alone.

Sulphur—Before the fit: creeping and crawling as of a mouse running down the back and arms, or as if a mouse were running up the leg. Chronic cases. Eruptions on the skin, or may have been suppressed.

Consult also the remedies under Convulsions.

---

4. Convulsions of Children, (*Eclampsia.*)

Due principally to changes in the brain; teething; some irritating substance in the stomach or bowels, as worms, indigestible food, &c.

TREATMENT—If due to something the child has eaten an emetic may expel it and relieve the whole trouble. Bathe the feet in water hot as can be borne. Immersing the child in hot water for a few moments, generally relaxes the spasms.

Aconite—High fever. Hot, dry skin. Hiccough. Restlessness. Moaning. Child gnaws his fist. Twitching of single muscles. Jerks the left leg or arm.

Aethus—Clenched thumbs. Eyes turned down. Pupils enlarged, immovable. Teeth set. Sudden violent vomiting of frothy, white substance; or yellow bile; or curdled milk. Diarrhea of yellow or greenish mucus with vomiting of green slime. Face swollen, with red spots. A line or fold from each corner of mouth to nose.

Arsenicum—Great restlessness before spasms. Thirst, drinks little and often. Patient lies motionless, then the mouth is drawn first to one side and then to the other, a violent jerk seems to pass through the whole body and conciousness gradually returns.

Belladonna—Head hot. Face flushed. Eyes red and staring. Pupils dilated. Drowzy but can't sleep. Starts and jumps during sleep. Eye balls twitch and roll. Mouth twitches. Arteries in the sides of neck throb.

Chamomilla—During sleep works the mouth as if smiling. Hot sweat on forehead and in hair. One cheek red. The child is very cross, wants to be carried all the time; wants various things which are refused when offered. Stiffens and bends backward. Legs move up and down. Reaching and grasping with hands.

Cina—Becomes stiff suddenly; there is a clucking sound as though water was poured out of a bottle, from the throat down to the abdomen. Frequent swallowing as if something was in the throat. Worm symptoms. Child constantly works at his nose. Urine turns milky after standing.

Ipecac—Nausea and vomiting accompany spasms; Body stiff, and stretched out, followed by jerking of the arms.

Opium—Spasms from fright, anger, emotions &c., from approach of strangers. Throws or stretches the limbs at right angles with the body; bends backward and rolls sideways. Deep sleep after attack or between spasms. Pupils of eyes contracted. Spasms from nursing soon after the mother has been frightened.

5. St. Vitus' Dance. (*Chorea.*)

St. Vitus' dance consists in involuntary movements of the muscles or limbs, without loss of consciousness. Generally, any effort to control these movements makes them worse. Sometimes the jerkings will be confined to a single muscle or limb; sometimes they will begin in an extremity and extend over the whole body. Usually these movements cease during sleep; although they occur to some extent during dreams.

TREATMENT—Keep the body in good health, by proper regulations of bathing, food, clothing, exercise, amusement &c. If *secret habits* are indulged in try and break them up by plainly showing the evils they lead to. Do not *scold* the patient for the strange movements he cannot control, nor ridicule him. Reading that unduly excites the imagination should be prohibited. Nor should study be allowed if it excites the brain.

Agaricus—Jerkings of different muscles. Dancing like turning of the whole body. Twitching of eyeballs and lids. Worse on approach of thunder storm.

Belladonna—Throws the body forward and backward while lying. Bores the head into the pillow. Grating of the teeth. Sore throat. Numbness of fingers. After mental excitement.

Cimicifuga—Motions of left side mostly. Worse during menses.

Cina—Motions begin with a shriek. Affects the tongue and throat, and continues through the night. Headache; enlarged pupils; dark rings around the eyes. Ravenous appetite. Pain around navel. Worm symptoms. White around the mouth, rubs the nose, milky urine.

Cocculus—Motions of right arm and right leg;

they cease during sleep. Face puffy and bluish, symptoms of paralysis.

Hyoscyamus—Throwing the arms about. Misses what he reaches for. Constant falling of the head from side to side. Tottering gait. Very talkative or loss of speech. Silly; smiling; laughing at every thing told him.

Ignatia—When caused by fright or mental excitement. Worse after eating. Better lying on back.

Nux Vom—Feeling a numbness in affected part after much drugging.

Sepia—Motions of head and limbs. When talking, jerking of muscles of face. Stammering. Very restless. Eruptions like ring worms, every spring.

Sulphur—Chronic cases. After suppressed eruptions. Sensation of a mouse running up the limb.

Sticta—She cannot keep her feet to the ground; they jump and dance around in spite of her, unless held fast. When lying down her limbs feel light as feathers; as though they were floating in the air.

Consult also remedies under EPILEPSY *and* CON-VULSIONS.

---

6. Paralysis.

Paralysis is a loss of the power of motion, or of both *motion* and *feeling*. It may be confined to a single muscle or limb, or it may affect the whole body. Its causes are changes in the brain or spinal cord. It frequently follows an attack of apoplexy. There is a peculiar prickling, crawling sensation as if the limb were asleep, at first; afterward numbness and heaviness.

TREATMENT —Electricity is said to have done good. It needs to be applied by one skilled in its use, or it will do harm instead of good. For remedies look over the sections on apoplexy, epilepsy, convulsions, &c.

Belladonna—Apoplexy. Congestion to the head. Paralysis of one side, spasms of the other. Paralysis of the face.

Gelseminum—Loss of motion but not of sensation. Difficulty of swallowing. Loss of voice after diphtheria.

Lachesis—Left side. Awkward, stumbling gait. Lifts the feet high when walking. After apoplexy.

Nux Vom—Partial paralysis of face, arms or legs, with dizziness. Darkness before the eyes. Weak memory. From sexual excesses; hard drinking; poisoning by lead, arsenic &c.

Opium—After apoplexy. In drunkards; old people. Stool and urine retained.

Phosphorus—Spasms on the paralyzed side. Crawling feeling in the limbs. Loss of feeling. Shuffling gait; hardly lifts the feet.

Plumbum—Paralysis preceded by mental derangement, trembling, spasms, or shooting, tearing pains. Paralyzed parts become smaller. Paralysis alternates with colic.

Stramonium—After apoplexy. Left side. Stammering. Sheds tears. Paralysis of one, spasms of the other side.

7. Neuralgia of Head and Face.

(*Prosopalgia, Tic douloureux.*)

Neuralgia is recognized by very severe pains

coming in paroxysms. Affects generally but one side at once.

TREATMENT—*Aconite*—Cheeks red and hot. Patient seems beside himself for pain. Screams and rolls about in bed or on floor. Dizzy on rising from sitting position. Fear of death.

Arsenicum—Attacks return periodically. Pains burning, stinging, like red hot needles. Great fear and restlessness. Worse about midnight. Better from warmth, and moving about. Very weak. Thirst for little and often.

Belladonna—Right side mostly affected. Pains worse under right eye, increased by rubbing or touching the part. Shooting pains in eyeball. Muscles of face twitch. Face flushed. Worse from noise and light; and in afternoon. Pains commence under left eye and run back to ear.

Chamomilla—Stitching, jerking pain, worse at night. Hot sweat about the head and in the hair. Crying out from pain. Patient very uncivil. Cross.

Colocynthis—Tearing, darting pains on left side mostly; worse from least touch. Better from rest and warmth.

Gelseminum—Beating pains. Eyes affected. Dim vision. Upper eyelids heavy, can hardly keep eyes open. Twitching and drawing of muscles near part affected. Mind confused.

Hepar—Chronic cases. Pain shoots from cheek bone to the temple, ear, wing of nose and upper lip. Worse in fresh air.

Lachesis—Left side. Pain in and about the eye. Heat rises to the face before the attack, and weak, sick feeling in the abdomen afterward.

Mercurius—Tearing pains, worse at night in bed. Often starts from decayed tooth. Great quantity of saliva in mouth. Sweats much. From taking cold.

Nux Vom—Tearing pains. Eyes red and water. Flow of clear water from nose. Numbness of affected part. Patient is morose, irritable; belches a great deal; is constipated. After abuse of coffee, liquor, quinine &c.

Phosphorus—Drawing and tearing pains in the jaws, at the root of the nose, in the eyes and temples, with bloated face, dizziness, and ringing in the ears. From taking cold over the wash-tub.

Pulsatilla—Twitching, tearing pains, worse in the evening and in a warm room. Scanty or suppressed menses from getting feet wet. In persons of mild, tearful disposition.

Rhus tox—Pain is burning, tearing or drawing. Teeth feel too long. Worse after midnight; and worse when quiet, must move continually to get a little relief.

Sepia—Pains intermit, congestion of head and eyes. Jerking, like electric shocks upward. Mostly on left side. Yellow face; yellow across nose.

Sanguinaria—Pain in upper jaw extends to nose, eye, ear, neck and side of head; shooting, burning pains; must kneel down and press head tight to floor.

———

8. Headache. *(Cephalalgia Migræna.)*

Apt to occur periodically. Caused by bad air; lack of out-door exercise. Indigestion. No doubt the use of tobacco, tea and coffee frequently causes it. *Tight lacing a prolific cause.*

TREATMENT—When the case has become chronic it will require perseverance to effect a cure. All causes should be removed. Tobacco, tea and coffee should be given up, or but sparingly used.

Follow advice found in GENERAL DIRECTIONS and under TREATMENT in CHRONIC CATARRH. The skin, stomach and bowels must be kept in good condition. Remove every hindrance to the free circulation of the blood in the extremities.

Aconite—Violent pain, with great fulness and heaviness in forehead as if the brain would be pressed through the eyes. Crampy pain over root of nose. Dizzy and faint when rising up. Bitter, bilious vomiting. Restlessness. Fear of death.

Aethusa—Head feels as if in a hoop. Headache better from passing wind downwards. Pain in the forehead as if it would split, at its height vomiting and finally diarrhea. Eyes protrude, pale face. Better in the open air.

Arsenic—Beating ; or pressure as from a load on the brain. Rising up in bed or moving makes it worse ; cold washing relieves temporarily, walking in open air relieves permanently. Pain in head and face is worse on left side. Violent vomiting particularly after eating and drinking. Great restlessness, weakness, fear of death. Drinks little and often. Feels chilly, hovers near stove. Headache alternates with bilious colic.

Belladonna—Fulness and throbbing in brain. Pain in head and eye-balls, eyes feel as if starting from their sockets. Face flushed and eyes red. Pain obliges one to close the eyes. Throbbing in arteries of the neck. Pain begins suddenly and ends suddenly. Worse from noise, jar, bright light, and about 3 P. M. Worse leaning forward, or lying down. Better tying up the head or pressing the forehead. Right side usually. *Belladonna* is useful in most cases where there is a feeling of fulness or congestion of the head.

Bryonia—Headache begins generally on first waking in the morning, gradually increasing, worse from coughing, sneezing or any motion. Pain as though head would burst. When bending forward feels as if brains would fall through forehead. Wants to keep perfectly still. Sour vomiting, or bitter. Thickly coated tongue. Constipation, stools hard and dry. Patient very irritable. Left side mostly.

Calcarea carb—Chronic headache. Head feels icy cold ; generally on side. Throbbing headache in middle of brain every morning, lasts all day. Headache from over-lifting. Worse going up stairs, talking or walking ; in hot sun. Better from bandaging ; pressure of cold hand. Feet constantly cold and sweaty. Menses (monthly sickness) too soon, too profuse and lasting too long.

Calcarea phos—Headache of school girls who study much. Of young persons who grow fast. Head aches on top and back of ears. Crawling feeling on top. In those who suffer from grief or disappointed love.

China—Pain worse from slight touch. Ringing and roaring in ears. Stitches from temple to temple. Brain feels bruised. Scalp very sensitive. Worse every other day. In nursing women, or those who are weak from loss of blood or other fluids.

Chamomilla—Patient very irritable and impatient. Shooting or throbbing pain in forehead. One cheek red the other pale. Bitter vomiting.

Cocculus—Headache from riding in a carriage or boat or on the cars. Worse after eating and drinking. Head feels empty. Pain in back of head and neck as if opening and shutting like a door.

Colocynthis—Tearing pain through head, worse from even moving upper eye-lid. Sweat smells like urine. Urine scanty. Headache worse from anger.

Glonoine—Sensation of fulness and throbbing goes from chest to head. Head feels big. Before headache there may be tightness of the chest and difficult breathing. Feeling like the motion of waves in brain. Eyes red.

Gelseminum—Head feels heavy, big. Sensation of a band around the head above the ears. Dim sight, eye-lids heavy. Hunger before and during. Feels as if scalp were drawn tight over head.

Nux vomica—Headache usually worse in morning ; or begins in morning, abates toward evening ; stoppage of nose, sour or bitter vomiting. Constipation with frequent urging for stool without success. Persons who are indoors much ; who drink much coffee, or who are troubled with piles.

Phosphoric acid—After long continued grief. Headache of school girls Worse from least shaking or noise, especially music. Menses too early and too long. Great heaviness of the head.

Platina—Crampy pain above the root of the nose as though the part were in a vice. Numb feeling in the brain. Menses too early, too profuse and too short lasting. Hysterical.

Podophyllum—Stunning pain through temples, better from pressure. Mist before the eyes, then fleeting pains in back of head and down neck and shoulders, better when lying in a quiet and dark place, and from sleep. Darts through the forehead obliging one to shut the eyes. With bilious symptoms. Yellowness of the skin and eyes. Dull pain in right side. Sour or bitter risings from stomach.

Phosphorus—Headache over left eye ; in left side of head. Every other day. Burning and beating in forehead, mostly in forenoon, with nausea and vomiting until noon; worse from music, chewing, and in a warm room. Females who suffer from grief. Tall slim persons with dark hair and eyes.

Sanguinaria—Pain generally begins in back of head and passes upward and forward and settles above right eye; nausea, vomiting and chilliness. Wants to lie still in a dark room. Gets relief by lying with back of neck on something hard. Better from sleep; from vomiting. *Sanguinaria* will relieve most cases of ordinary sick headache.

Pulsatilla—Jerking, tearing, stitching pains, worse evening and night. Pain in the side of head lain upon; when turning on the other side the pain goes to that side; worse evening and looking upward. Headache worse when quiet. No appetite nor thirst. Water tastes bitter. In persons of mild yielding disposition.

Sepia—Stinging, shooting pains, from within out, mostly in left side and over left eye, with vomiting and contraction of pupils of eyes : better lying on painful side. Throbbing headache in the back of head, beginning in the morning, lasting till noon or evening; worse from least motion, turning the eyes, lying on the back ; better lying on side, closing eyes, at rest in dark room. Pale, yellowish, sickly color of face. Yellow spots on face. Yellowish across nose. Leucorrhœa. Bearing down pains in abdomen. Constipation.

Sulphur—Sensation of heat in top of head. Burning in soles of feet at night or coldness of feet. Gets faint and hungry at 10 or 11 in the forenoon. Bowels loose early in the morning. Eruptions on skin. Troubled with piles. Lean, stoop-shouldered persons. Throbbing headache every night.

Cimicifuga—Great pains in head and eye-balls. Feels as if top of head would fly off. Pain just back of eye-balls. Head feels heavy and big. Dizzy. Declares she will go crazy.

See also Biliousness.

---

1. Falling of the Hair. (*Alopecia.*)

Falling of the hair follows fevers, chronic head-aches, &c. Caused also by dyes, invigorators, renewers, etc.

TREATMENT—*Hepar*—Falling of the hair with very sore, painful pimples and large bald spots on scalp. Hard lumps on head, sore to touch.

Phosphorus—Roots of hair get grey and hair comes out in bunches. Dandruff copious. Itching after scratching.

Sepia—Roots of hair sensitive, worse in evening, from touch, cold north winds, and when lying on the painless side. Burning after scratching.

Sulphur—Hair dry; falls off. Scalp sore to touch, itching violently in the evening, when getting warm in bed.

See also Headache.

---

2. Scald Head. (*Eczema capitis.*)

This consists of an eruption on the scalp, of a great number of small vesicles. These dry up and form thin scales, or break and discharge a fluid which dries into a scab or crust. This disease is more disgusting than dangerous. Quite often some other ailment will cease when this appears. Nature is trying in this way to relieve the system of some virus, and it is a great mistake to attempt to cure this disease by external applications. The crust should be thoroughly softened by pure, sweet lard before any effort is made to remove it. In fact, it is often best to let the crust remain until the surface beneath it is entirely healed.

TREATMENT—*Calcarea carb*--Thick, creamy discharge. In children who have large stomachs, light hair and sweat much about the head; who get their teeth slowly. Glands of neck swollen.

Lycopodium--Eruptions begin on back of head; crusts are thick. Pus is thick and mild. Child is cross on awaking.

Hepar--Moist eruption of bad odor; Boils on head and neck. Sour smelling stools.

Clematis--Eruption on back of head extending down the neck; itching worse after getting warm in bed. Worse washing in cold water; worse during increasing moon, better during decreasing moon.

Graphites--Large, dirty crusts which mat the hair together. The moisture oozing out sticky like glue.

Petroleum--Moist eruption, most on back of head; also behind ears.

Rhus tox--Eruption forms thick crusts; itching worse at night. Eats the hair off; eruption extends to shoulders.

Sulphur--Dry, scabby eruption begins on back of head, burning and bleeds easily; cracks. Itching worse when warm in bed.

Viola tri--Thick yellow matter glues the hair together. Thick crusts. Often passes urine involuntarily. Milk crust in children lately weaned, with violent cough and oppression of chest. Child twitches with his hands in his sleep, with his thumbs clenched.

3. St. Anthony's Fire. (*Erysipelas.*)

May attack any part of the body, but most often the scalp and face. It is an inflammation of the skin. It is preceded for a day or two by chilliness and feverishness and general bad feelings; then the part affected becomes red, hot and painful. The skin swells and becomes very sensitive to the touch. Blisters often form. The inflammation may extend to the brain; the patient becoming unconscious and delirious. A wound or scratch may induce it. Intemperance, uncleanliness, and foul air predispose to it.

TREATMENT—*Belladonna*—Right side. Skin red, hot and shining; Delirium. Pupils dilated. Cannot bear light or noise. Worse 3 or 4 in afternoon.

Cantharis—Burning, stinging pain. Begins on dorsum of nose, spreads to both cheeks, more on the right. Skin peels off. Formation of blister. Passes little or no urine. Frequent effort to pass urine with burning and smarting.

Rhus tox—When on left side spreading to right. Blisters form. Burning, tingling and itching in part affected. Face dark red, covered with yellow vesicles. Restlessness. Sleeps best after midnight.

---

4. Milk Crust. (*Crusta Lactea.*)

This is a disease of nursing infants. Confined usually to the forehead and cheeks, yet may attack the whole body. Commences as a patch of pimples. This patch becomes more inflamed and spreads by the constant rubbing and scratching. The discharge becomes milky or yellow, drying into scabs and crusts.

TREATMENT—Prevent the eruption being rubbed

or scratched, even if the hands of the child have to be secured. The best application to relieve the itching is a mixture of lime water and sweet oil, which may be prepared as follows: Shake a little lime in pure soft water for a few minutes; allow the lime to settle; mix together equal parts of the clear water and olive oil; apply with a feather or soft brush. Use the utmost gentleness in washing the child's face

Arsenicum—Discharge which seems to make sound places sore. Itching and burning worse at night, worse in cold air, better from warmth.

Calcarea carb—Children with light complexion, light hair, fat. Sweat much about head. Large stomach. Sweats after eating or drinking. Worse about new moon; after washing.

Graphites—Eruption discharges a watery, sticky substance. Cracks and chaps on cheeks. Soreness behind the ears, with sticky moisture. Constipation, stools large, knotty.

Lycopodium—Thick crusts, surface raw and cracked underneath. Raw places on the skin, in the groin. Much wind in stomach and bowels. Child seems hungry but a few swallows seems to fill him up. Belching of wind, can hardly swallow.

Rhus tox—Glands about the neck and throat swollen. Stiff neck.

Sulphur—Violent itching, worse at night. Bleeds after scratching. Diarrhea in the morning.

Viola tri—*See under Scald Head.*

---

1. Sore Eyes. (*Opthalmia.*)

The symptoms are too well known to need describing. *Causes* are various; foreign bodies, sparks, cinders, pieces of coal, steel &c; sharp winds; too strong light; using the eyes too much; intemperance. A scrofulous constitution predisposes one to this disease. *Opthalmia* often occurs in new-born infants.

TREATMENT—Remove with the utmost gentleness any foreign bodies. For this purpose use a soft camel's hair-brush or soft cloth, or the loop of a horse hair doubled. Keep the eyes clean as possible with tepid warm water and castile soap. In extremely bad cases the patient must remain in a darkened room. A filthy and hurtful practice is that of applying the mother's or nurse's milk to the sore eyes of infants; the milk soon becomes sour and irritates the eyes.

Aconite—In the outset. Eyes dry, hot, red. Upper half of eye-ball sore. From cinders or other foreign bodies. High fever, hot, dry skin. Restlessness.

Argentum nit—Inner corner red and swollen, looks like a lump of red flesh. Clusters of red vessels extend from inner corner toward the pupil.

Belladonna—Sudden attack worse in right eye. Eye ball red and prominent. Shooting pains; Throbbing pains. Cannot bear light or noise. Pains appear and disappear suddenly. Throbbing headache worse from motion.

Arsenicum—Burning pains. Eruption on the face, and under the eyes places made sore by the discharge. Very weak. Drinks little but often.

Euphrasia—Sensation of a hair hanging over the eye and must be wiped away. Profuse flow of tears which are acrid and make the parts sore. Cold in the head with profuse discharge of bland mucus. Violent itching of eyelids.

Hepar--Little pimples surround the inflamed eyes

Mercurius—Eye very sensitive to lamplight or glare of fire. Cold in the head. Nose sore from the discharge. Upper lip swollen.

Rhus tox—Profuse flow of acrid tears in the morning and in the open air. The cheek under the eye is dotted with red pimples. The lids are kept tightly closed, when opened a great gush of tears.

Sulphur—Chronic. Edges of lids thickened and ulcerated. Stitches through the eye into the brain, worse in sultry weather. Eruptions on the body. Heat on top of head. Burning of soles at night. Faint, gone feeling at 10 or 11 A. M. Early morning diarrhea. Sensation of a hair in the throat.

Pulsatilla—Child frequently rubs the eyes. Discharge thick, yellow, profuse, glues the lids together. Styes, especially on upper lid.

----

2. Inflammation of the Ear (*Otitis,*) **Earache** (*Otalgia,*) **Discharge from the Ear,** (*Otorrhœa.*)

Inflammation may result from taking cold; from syringing the ears too forcibly, or from violence in the use of pins, needles, hair-pins, ear-spoons &c. It is an exceedingly painful disease. Little children carry their hands to the painful ear, throw their heads from side to side, bore it into the pillow, scream and become more uneasy when rocked. If nursing they suddenly let go the nipple, because the effort to suck increases the pain. *Otorrhœa* follows previous inflammation. It very frequently occurs after Scarlet Fever.

TREATMENT—Hot fomentations, steaming sometimes affords relief.

Aconite—Very sensitive to noise. Tearing in left ear. Ear is red, hot, swollen and very painful and sore. Chilliness or feverishness. Hot dry skin. Thirst.

Aethusa—Yellow discharge from right ear, with stitching pains.

Bryonia—Swelling, redness, heat and soreness of right ear-; at times piercing stitches deep into the ear; pain and swelling of the gland under the ear.

Belladonna—For same symptoms, with throbbing headache, beating in the arteries on the sides of the neck; throat sore and dry. Pains seem to come and go suddenly. Red face, head hot.

Hepar—Hardness of hearing. Cracking in ear when blowing the nose. Bad smelling discharge from the ears. Itching of outer ear. Scurfs on and behind the ears. Ear very sensitive to touch. Can't bear to have it get cold. After scarlet fever.

Kali bi—Violent stitches in left ear, extending to roof of mouth, side of head and neck. Thick, yellow, bad smelling discharge from ears. After scarlet fever.

Mercurius—External and internal ear inflamed, with ·tearing, stinging pains. Green, offensive discharge. Glands swollen. Canker sores in mouth; great flow of saliva. Sweats at night, but no relief from it.

Pulsatilla—Pain in ear through the night, worse in paroxysms; little suffering during day. Discharge mild and not very offensive.

Sanguinaria—Earache with headache, singing in the ears and dizziness. Right ear worse.

Sulphur—Stinging in left ear. Offensive discharge, worse left ear. Discharge about every eighth day. Humming or hissing in the ears.

3. Deafness.

May be due to a variety of causes. Generally incurable.

TREATMENT—Hardened ear wax should be removed by filling the ear with sweet oil and letting it remain over night; in the morning syringe out the ear with warm water and castile soap. This can be repeated until the ear is clear.

Mercurius—Sounds vibrate in the ears. Hears better for a moment after swallowing or blowing the nose. Constant cold sensation in the ears.

Phosphorus—Difficult to hear the human voice. Noises in the ear. Sounds re-echo in the ears, especially music.

Pulsatilla—Deafness after measles. From taking cold after having hair cut. Hard, black wax. Can hear better on the cars.

––––

1. Cold in the Head. *(Coryza   Catarrh.)*

This well known complaint is an inflammation of the mucous membrane that lines the cavities of the nose. It sometimes extends to the eyes, throat and air passages. It is generally caused by taking cold, or inhaling some irritating substances. It usually begins with sneezing and stuffed up feeling of the nose; following this a discharge of mucus. Cough and hardness of hearing often accompany it.

CHRONIC CATARRH. *(Ozaena.)*—Is generally the result of repeated attacks of simple *Coryza*. An offensive discharge is expelled from the nose and trickles backward and downward into the throat. The voice is changed and permanent deafness is sometimes the result.

TREATMENT—Simple *Coryza* yields readily to the appropriate remedy. *Chronic Catarrh* is always accompanied by a disordered condition of the *skin* and *circulation*. The body should be thoroughly bathed once or twice a week ; and the *flesh brush* used *vigorously* every day. Every thing that hinders the free circulation of the blood should be removed. Corsets, elastics, belts &c., should not be worn ; the clothing should be loose and supported from the shoulders. *Breathing should be done through the nose and not through the mouth.* Flannel underclothing should be worn the entire year. The feet should be kept dry and warm. A bad practice is that of leaving a warm room for a cold bed on retiring. This is particularly trying for the enfeebled or the aged. Avoid all snuffs, douches, and so called *cures*. They may *tan* and *thicken* the mucous membranes, but this is not a *cure*. And I am confident that *Consumption* has more than once started from a *suppressed* catarrh. An adherence to the above directions with the proper Homeopathic remedy will cure every curable case. Bad cases should receive professional treatment.

Aconite—In the beginning, dry state. Sneezing, fever, thirst, restlessness, headache, roaring in the ears. Better in the open air.

Arsenicum—Burning, watery discharge, causing soreness and smarting of nostrils. Stoppage at bridge of nose.

Belladonna—Fluent discharge from one nostril only. When blowing the nose, smell of herring brine. Headache, worse on motion. Feeling of great fulness in forehead.

Cepa—Constant sneezing, with profuse acrid discharges when coming into a warm room. Tingling and itching in nostril. Headache, cough, thirst, want of appetite, trembling of hands; worse evenings. Better out of doors. Sensitive to odor of flowers and skins of peaches.

Euphrasia—Profuse discharge of white, mild mucus. Eyes inflamed and full of acrid tears. Cough only through the day.

Hepar—Catarrh, with inflammatory swelling of the nose as from a boil.

Kali bi chrom—Ropy, tough discharge. Discharge of hard green masses, or hard plugs. Pressure and tightness at root of nose; worse evenings and in open air. In the morning obstruction and bleeding from right nostril. Tickling as from a hair up in the nostril.

Mercurius—Dropping of water from nostrils. Nose swollen, red and sore. After sweating in the night the cold is no better in the morning. Feels bad in a warm room, neither can he bear the cold.

Phytolacca—Flow from one nostril, the other stopped, both stopped while riding.

Pulsatilla—Thick, yellow, green discharge. Loss of smell and taste. No appetite nor thirst. Feels better in fresh air; worse in warm room. Chilliness.

Rhus tox—Thick, yellow, or green discharges. Fever-blisters and scabs under the nose. Tip of nose red and sore; sore inside. Aching in all the bones. Feels bad when at rest.

Sulphur—Profuse discharge of burning water; flows freely out-doors; nose stopped up in-doors. Chronic stoppage; also of one nostril.

Rose Cold, Hay Fever, Hay Asthma—A catarrhal difficulty that attacks some persons at the same date every year, generally in the summer or fall. Symptoms are similar to those of a common cold in the head, only that it is very obstinate, lasting for four or six weeks. Remedies preceding may be of use.

SNUFFLES, OR STUFFED NOSE OF INFANTS—Principal remedies, *Nux vom*, *Sambucus*.

---

2. Thrush. Canker Sore Mouth. *(Apthae.)*

Although two distinct diseases they are usually described together. They are too well known to need description. The latter seems to be more frequently developed in scrofulous children.

TREATMENT—Always wash the mouth of the infant, and the nipple of the mother after nursing, with water or wine and water mixed.

Arsenic—Sore mouth in adults. Great burning and prostration. Worse at night, midnight.

Lachesis—Canker sores on tip of tongue.

Lycopodium—Canker sores under the tongue, near the cord.

Mercurius—Canker sores. Much saliva. Bad smell from mouth. Sweats at night, but no better for it.

Sulphur—Gums swollen. Bloody saliva runs from mouth. Sour smell from mouth. Stools slimy. Sore about the anus. Morning diarrhea.

---

3. Toothache.

TREATMENT—*Aconite*—From cold; from cold, dry winds, with throbbing on one side, redness of the cheek, congestion to the head, great restlessness. Left side, or from right to left. Teeth sensitive to cold air. Toothache in sound teeth.

Belladonna—Pain some minutes after eating, not during. Pain increases gradually to high degree and as gradually diminishes, or may begin, and cease suddenly. Teeth feel on edge. Dull pain in right side, upper row, all night. Painful swelling of gums on right side.

Bryonia—Teeth feel too long. Pain relieved by cold water; worse from anything warm. Tearing, stitching toothache while eating, extending to neck, worse from warmth. Jerking pain when smoking.

Hepar—Toothache worse in a warm room and when biting teeth together. Throbbing as if blood entered the tooth.

Mercurius—Pain in several teeth at once; extends to ear. From damp weather, evening air. Worse from eating or drinking anything cold. May cease during night to return next day De-cayed teeth. Better from rubbing cheek.

Pulsatilla—Toothache worse in a warm room; in a warm bed; from anything warm in the mouth. Better from cold; in the open air, walking about. Confined mostly to one side with earache. Drawing, jerking as if nerve was put on the stretch and then let loose.

Arsenicum—Jerking toothache, extending to the temple; relieved or removed by sitting up in bed. Relieved by heat of stove.

---

2. Disorders of Teething.

TREATMENT—Many times the disorders of this period are due to *over-feeding*. If there be vomiting or diarrhea, withhold all food for 6, 8 or 12 hours, then begin with the quantity lessened to an amount that the stomach will retain. A change of food may be necessary. See sections on *Dyspepsia Diarrhea Cholera Infantum.*

Aconite—Restlessness. Child gnaws at his fists or fingers. Great thirst; Skin hot, and dry.

Arsenicum—Restless after midnight. Child strikes its head or face. Vomits all fluids soon after swallowing. Drinks little and often. Gum over the tooth looks as if blistered, or filled with a dark watery fluid. Scaly eruptions on the scalp. Child has a pale, waxy look.

Aethusa—Face pale, or swollen, with red spots. Vomits curdled milk, or frothy white substance. Vomits green slime, and stools of same. Eyes turned downward. Canker sores in mouth.

Belladonna—Child moans. Starts and jumps both sleeping and waking. Throws up its hands. Awakens from sleep with a start and fright and staring eyes. Skin leaves a burning sensation to the hand. Hot sweat. Delirium. Shivers during stool. Stool involuntary. Worse afternoons.

Bryonia—Lips parched. Vomits when raised upright. Wants to lie perfectly still. Thirsty for much at a time.

Calcarea carb—Sweats about the head when asleep. Stools large, hard like chalk, or thin and white. Teeth long time in coming. Large stomach. Feet cold and damp. Swelling around the neck.

Chamomilla—Child must be carried all the time when awake. One cheek red. Stools look like chopped eggs and spinach. Thirsty.

Cina—Child works at his nose. Grinds his teeth at night. Unnatural appetite. Hacking cough followed with an effort to swallow something. Very peevish. Does not want to be spoken to, looked at or touched. Urine turns milky after standing. Head falls from side to side. Frequent swallowing.

Hepar—White, sour smelling diarrhea. Dry eruption on some part of the body. Loose cough, with choking.

Ipecac—Face pale, blue around the mouth and eyes. Constant nausea and vomiting. Spasms with stiffening backward.

Lycopodium—Child very cross when it awakes. Cries and screams before making water. Urine has a red sand or stains diaper red. Little food seems to fill him up. Worse from 4 to 8 P. M.

Mercurius—Much saliva runs from mouth. Ulcers and blisters in the mouth. Urine stains yellow. Stools slimy, bloody, with straining. Worse from evening until 3 in the morning.

Podophyllum—Grinding of teeth. Gagging. Painless diarrhea. Stools leave a meal like sediment on the diaper. Lies with half closed eyes. Great desire to press the teeth together.

Sulphur—Great soreness and redness of the anus. Eruptions on the skin. Child very averse to being washed.

Veratrum alb—Great nausea and vomiting. Great weakness after each stool. Cold sweat on forehead. Thirst for cold drinks.

Consult remedies under Convulsions.

---

1. Mumps. (*Parotitis.*)

Generally attacks children and young persons. It begins as a swelling below and in front of the ear, and sometimes extends over the whole side of the face. Ordinarily not serious, unless by bad treatment the gland becomes permanently hardened, or matter forms. Sometimes the testicles of the male, or the breasts of the female become affected.

TREATMENT—Avoid taking cold. Use no external applications unless it be dry heat Warm moist applications promote the formation of pus.

Aconite—May be used in the outset if there is chilliness, fever, thirst.

Belladonna—Bright red swelling; right side.

Mercurius—Pale swelling; right side Sweats at night. Much saliva.

Pulsatilla—When the breasts or testicles become affected.

Rhus tox—Swelling dark red ; left side.

Carbo veg. and *Arsenicum*—Testicles affected.

---

1. Quinsy, Sore Throat, *(Tonsilitis.)*

An inflammation of the tonsils. There is redness and swelling. Some fever usually attends.

TREATMENT—Simple external applications sometimes seem to do good. A mustard poultice is the best.

Belladonna—Right side worse ; parts bright red. Neck swollen outside, painful to touch and motion. Headache. Worse when swallowing liquids.

Hepar—Sensation as if a splinter or fish bone was sticking in the throat. Stitches extend to the ear when swallowing food.

Lachesis—Choking when drinking ; fluids are driven out through the nose. Throat feels constricted. Can't bear anything to touch the neck. Spells of suffocation, worse during or just after awaking from sleep. Worse, or begins on left side. Pain shoots into the left ear when swallowing.

Lycopodium—Begins on right side, goes to left. Throat dark red. Worse from warm drinks and after sleep.

Mercurius—Dark redness of throat. Much saliva. Bad odor from mouth. Canker sores. Glands on outside of neck swollen. Stinging, burning pains, worse from empty swallowing.

Kali bichrom—Shooting pains in left side, extend to the ear; relieved by swallowing. Tough stringy discharge.

———

2. Croup.

Is an inflammation of the upper part of the air passages. A membrane is formed and covers these parts. The child may or may not have for a day or two symptoms of an ordinary cold. It is aroused out of sleep, generally about midnight, by a hoarse, dry, barking cough. It soon falls asleep to be again awakened by the same cough. During the day the child may be lively and playful. Toward night the child becomes worse again; the cough more severe and the breathing more difficult.

TREATMENT—Apply no poultices, wet cloths &c., to the chest of the child. The breathing is already difficult enough. A flannel cloth folded 3 or 4 times may be wrung out of water hot as can be borne and laid upon the throat, and repeated as often as it becomes cold. Hot foot-baths may be given.

Aconite—Child awakens in his first sleep; very impatient, tosses about. Dry, short cough. Every *expiration* ends with a cough. Skin hot and dry. Child grasps its throat every time it coughs. *Aconite* is usually given at the beginning of any case, and often used in alternation with other remedies.

Belladonna—Wakens about 11, P. M. with a barking cough. Face fiery red. Cries before and during the cough. Cough ends with sneezing. Eyes red, pupils dilated. Starts and jumps during sleep.

Hepar—Loose cough, with choking as from phlegm. Child cries when coughing. Sneezing after cough.

Kali bichrom—The disease has approached gradually. Cough is · hoarse, metallic, like coughing through a brass tube. A membrane forms. Tough, stringy saliva in mouth and throat. Violent wheezing and rattling. In fat, chubby, light-haired children.

Sanguinaria—Wheezing, whistling cough. Circumscribed red spots on one or both cheeks.

Spongia—Dry, crowing, barking cough. Fits of suffocation, must throw the head back. Breathing· sounds like sawing through a board. Coughs at the end of *inspiration*.

---

3. Diphtheria: (*Diphtheritis.*)

This is a disease which shows itself mostly in the throat, mouth and nasal passages. It is marked by an inflamation and formation of false membrane in those parts. It may begin as a simple sore throat; the patient soon becomes very weak ; a bad odor is given from the parts affected ; bloody saliva may run from the mouth ; patches of membrane may be seen back in the throat; swallowing is difficult or impossible ; the false membrane may extend up into the nose or down into the air passages, in the latter case the cough becomes croupy. This disease is contagious.

TREATMENT—The temperature should be kept about 70 degrees, and the air moist by steam. Let the patient gargle his throat with brandy, or mixture of alcohol and water. This should be held in the mouth and throat until smarting is produced· If the patient cannot gargle he should inhale the vapor of alcohol frequently. Patches of membrane may be touched with a brush or swab wet with alcohol. The mouth and nostrils should be kept free as possible from the discharges by tenderly wiping with a moist rag, which should be burned afterward. Give the patient the most nourishing food he will take. A little of the *carbolate of lime* or *chloride of lime* may be sprinkled about the room ; or cloths dipped in a solution of *carbolic acid*, or better still, *salicylic acid*, may be hung about.

Belladonna—May be given in the beginning. Patient very restless. Sore throat. Pupils enlarged. Drowsy yet unable to sleep. Starts suddenly out of sleep. Head hot and painful. Will not lie down for fear of choking. Eyes red. Face flushed. Throbbing in sibes of neck.

Kali bichrom—Ulcers in mouth and throat. Mucus or slime streaked with blood. Tough, stringy mucus that may be drawn out in long strings.

Lachesis—Stinging pain extends to ear Wants to swallow, or to hawk up something, with choking spells. Dislike sto have the throat touched. Disease begins or is worse on left side. Worse after sleep.

Phytolacca—Chills the first days of the disease. Violent pain in front or back part of head, in the back and limbs. Choking sensation. Membrane has a gray color generally.

Kali permanganate —Has cured some very bad cases. Should use it if other remedies fail. Dissolve one grain in a glass of water and give a teaspoontul every hour.

---

1. Bronchitis. Inflammation of the Lungs.
(*Pnuemonia*.) Pleurisy. (*Pleuritis.*)

BRONCHITIS--Consists in an inflammation of the mucus membrane or inner lining of the air tubes. Generally the result of taking cold. Begins usually with chills alternating with heat. Often there is a tickling or burning under the breast bone, with a stuffed up feeling. Cough, at first dry, becomes loose and rattling. Children often die from suffocation, not having strength to raise the phlegm.

INFLAMMATION OF THE LUNGS—Usually begins with a single hard chill followed by dry fever heat, and sweating. Shortness of breath ; cough ; tough, colorless phlegm may be raised at first, which may become rusty, bloody, yellow or green. Pain may be severe or slight.

PLEURISY—Begins˙ usually with a strong chill followed by high fever. Generally, pleurisy is accompained with a cough, difficult breathing, and violent, stitching, cutting pains in the sides of the chest.

TREATMENT—*Aconite*—In the beginning. Chill. Heat. Fever. Dry, hot skin. Thirst. Restlessness. Dry, hacking cough. Can't lie on right side. Sharp pain in left side. Fears he will die.

Belladonna—Great heat of skin yet sweaty. Crying before and when coughing. Sleepy, but can't sleep. Starting in sleep. Delirium. Red, hot face and head. Eyes staring. Pupils dilated.

Bryonia—Wants to lie still. Cough tight with stitching pains. Worse from motion ; during the day ; in a warm room. Pain in pit of stomach when coughing. Sharp stitching pains through chest when coughing or moving.

Mercurius—Stitching pain through to back when coughing, mostly on right side. Sweats without relief. Cannot bear either warm or cold air. Cough worse lying on right side. Thirst for ice-water, although it makes the cough worse. Bilious symptoms.

Ipecac—Rattling of large bubbles. Convulsive cough with throwing up of phlegm. Difficult breathing. Nausea. Vomiting. Face pale or bluish.

Kali bichrom—Tough phlegm, which can be drawn into long strings.

Phosphorus—Tight cough, worse before midnight. Tightness across the chest. Pain in the head when coughing. Rust-colored phlegm. Cough worse lying on back or left side. Stitches in left chest, better lying on right side. Water thrown up as soon as it becomes warm in the stomach.

Sanguinaria—Circumscribed redness of one or both cheeks. Burning in the ears. Burning in the chest. Lies on his back. Feet and hands burning hot; face and limbs cold. Breath and spittle smell badly, even to the patient. Belches before and after cough. Worse afternoons.

Tartar emet—Great rattling of phlegm, of which great quantities are raised; or none raised, with blue puffy face. Chest sounds as if full of mucus but none is raised.

See also section on cough.

------

2. Consumption. (*Tuberculosis, Phthisic.*)

A disease of the lungs accompanied with the formation of tubercles. These are round, greyish granules of the size of poppy seed. They may be seperate or in masses the size of a pea and upwards. These masses may soften into pus, be coughed up, forming abcesses and cavities in the lungs. This disease is most apt to show itself between the ages of twenty and thirty-five years. It is hereditary in most cases; and usually begins with a dry, hacking cough, easily excited by dust, active exercise, singing, loud laughing, exposure to cold air, &c. The person gets out of breath easily, and may suffer from various pains about the chest. After a time the cough becomes loose : yellow or greenish mucus, sometimes tinged with blood, may be raised. Toward the last, exhausting night sweats and diarrhea hasten the end. The patient often remains cheerful and hopeful to the last. The *medical* treatment of consumption should be entrusted to an intelligent Homeopathic Physician only. I am confident that a majority of the cases in the first stage, and many in the second, may be cured. A cough, troubling a person predisposed to consumption, should not be neglected.

---

3. Bleeding at the Lungs. *(Hemoptysis.)*

When the bleeding is from the very small vessels, the blood is coughed up; when, however, the larger blood-vessels burst, the blood gushes out through the mouth and nose in a stream. It may be caused by external injury, violent straining, &c. Usually results from some previous disease of the lungs, consumption, pneumonia, &c., or from the suppression of some habitual discharge of blood, as piles or menses.

TREATMENT—*Aconite*—Great restlessness and fear of death. Face expresses anxiety. Palpitation of the heart.

Belladonna—Cough from tickling in the throat. Fulness of the head. Face red and hot. Throbbing in the neck.

Hamamelis—Dark blood ; comes into the mouth without any effort, like a warm current from the chest. Mind calm. Taste of sulphur in the mouth.

Phosphorus—Spitting of blood in place of menses. Day, tight cough. Worse from evening till midnight.

Pulsatilla—Blood dark, clotted. Sad, weeping disposition. Chilliness. Loose stools. Menses suppressed.

Rhus tox—After straining, lifting, blowing of instruments; or when bleeding is renewed immediately after worry or mental excitement; bright blood.

———

4. Whooping Cough. *(Pertussis.)*

In the first stage like an ordinary cold. In the second stage the peculiar *whoop* is manifested. This is a loud *crowing* sound during the inspiration. The cough now occurs in paroxysms, ending generally with vomiting of tough, slimy phlegm. During the third stage the symptoms abate into those of an ordinary cold.

TREATMENT—*Coccus cacti*—Violent paroxysms end in vomiting of great quantity of thick, tough, stringy mucus.

Drosera—Paroxysms worse after midnight.

Cina—Child suddenly becomes stiff; there is a clucking sound like water poured out of a bottle, from the throat to the stomach. Child keeps swallowing. Worm symptoms. Urine milky.

Cuprum—Children get stiff with the whooping, breathing ceases, twitchings, after a while consciousness returns, they vomit and recover, but slowly.

Ipecac—Nose bleed, bleeding from the mouth, vomiting. Child looses breath, turns pale or blue and becomes stiff.

Veratrum—Spells brought on by entering a warm room or drinking cold water. Cold sweat on forehead and involuntary discharge of urine during coughing spells.

———

5. Asthma.

Pure and simple is supposed to be due to an irritation of the nerves that supply the lungs. There seems to be a constriction or narrowing of the smaller air tubes and vesicles. One can produce the sensation very nearly by breathing through a piece of thick cloth. The disease occurs in paroxysms generally; the intervals between may be of weeks duration. Changes in the temperature; dust; the odor of certain drugs will excite an attack.

TREATMENT—*Arsenic*—Worse at night. Must spring out of bed at night. Pale, cold face covered with cold perspiration. Restlessness. Worse in stormy weather and heavy air.

Ipecac—Chest seems full of phlegm yet none is coughed up. Cough causes vomiting which gives some relief. Violent constriction of throat and chest.

Lachesis—Worse after sleep; from touching the throat; after eating; moving the arms; better sitting up bent forward.

Dulcamara—Worse in damp, cold weather. Loose cough. With face ache after disappearance of eruption in the face.

Nux Vom—Asthma with fulness of stomach; better after belching. Tight sensation of lower part of chest, worse from cold air. Worse mornings.

Veratrum—Worse damp, cold weather; in early morning. Wants to move. Better throwing head back. Cold sweat on upper part of body.

Aconite—Face red, eyes staring. Fears he will die. Feels as if there was a band around the chest. Urine scanty, dark, pulse like a thread. After suppression of some rash.

Belladonna—Asthma in hot, damp weather; worse after sleep; worse in afternoon. Dilated pupils.

Cuprum—Attack of asthma begins with hiccough. Violent attacks come on suddenly, last some time and cease suddenly.

----

6. Cough.

This alone is a mere symptom, yet so annoying at times as to need direct treatment.

TREATMENT—*Aconite*—Short, dry, hacking cough. Worse from drinking cold water; tobacco smoke; lying on either side. Dry heat of body, thirst and restlessness. Child grasps its throat every time it coughs.

Belladonna—Cough ends with sneezing. Bloody taste in mouth. Pain in head or nape of neck when coughing. Child cries before the cough and during it. Barking cough wakens at 11 o'clock at night; face red and hot. Starts and jumps during sleep.

Bryonia—Cough, with stitching pain under breast-bone. Cough preceded by a crawling and tickling in pit of stomach. Worse after eating or drinking, or coming into a warm room. When coughing, feeling as if head and chest would fly in pieces.

Chamomilla —Paroxysms of coughing about midnight, when something seems to rise in the throat, as if it would suffocate. Tough, slimy phlegm tasting bitter. Children cough during sleep. One cheek red, the other pale.

Cina—Gagging cough. Cough followed immediately by swallowing. Urine turns white. Child bores and picks at his nose.

Hepar—Croupy cough with loose rattling and choking as of phlegm in the throat. Being uncovered or exposed to air excites the cough. After cough sneezing.

Kali bichrom—Cough with a pain from middle of breast-bone to back. Cough worse when undressing; after eating. Better after getting warm in bed. Phlegm, tough, stringy.

Mercurius—Violent cough, worse at night, as if head and chest would burst. Worse at night, in damp weather. Sweats at night without relief.

Nux Vom—Cough worse after midnight and in morning, and after eating; causing pain and soreness in stomach and bowels.

Phosphorus—Cough with tightness across the chest; trembling of the whole body; sticking in the pit of the stomach; stitches over one eye, splitting head ache, hoarseness, loss of voice; worse evening and night. Cough when any one enters the room; before a thunder storm; from strong odors; worse lying on back or left side. Cough in the tall, slender, rapidly growing; or tall people with dark hair and eyes.

Pulsatilla—Dry cough at night, going off when sitting up. Loose cough, worse at night, with diarrhea and vomiting of mucus. Coughs up dark clotted blood, menses suppressed.

Rhus tox—Dry, teasing cough, worse each evening and night until 12 o'clock. Stiffness in the

back and limbs. Cough excited by any part being uncovered, even a hand.

Sanguinaria—Wheezing, whistling, cough. Dry
cough awakens him and does not cease until he
passes wind upward and downward. Belching
before and after coughing. Gaping after cough.

Tart. emet—Loose, rattling cough, yet does not
raise. While coughing, sweat breaks out, gaps,
yawns. Nausea and vomiting of phlegm. Gasp-
ing for air before cough.

Lachesis—Gagging cough, persistent; from tickling in the throat, under the breast-bone, or in the
stomach; worse when falling asleep; from having
the throat touched.

Sepia—Cough with expectoration in the morning of yellow, green, gray or milk white mucus,
tasting salt. Cough commences every morning
before getting out of bed, lasts until 9 A. M., worse
when at rest; when lying on left side; from acids.

Sulphur—Cough with expectoration of greenish
lumps of a sweetish taste. When coughing, pain
as if chest would fly in pieces.

---

7. Palpitation of the Heart.

When merely of a nervous character it occurs in
spells, with generally entire freedom during the
intervals, which may be of weeks or months duration. *Palpitation* usually accompanies *organic*
changes of the heart; diseases, of which it is beyond the scope of this little work to treat, and
which should be confined to professional treatment. *Nervous palpitation* we find in persons whose
blood is deficient in quantity or quality; in women
at the change of life; in young persons who grow

very fast; in hysteric women; after sexual excesses, and from mental emotions.

TREATMENT—*Aconite*—Great fear of dying and restlessness.

Belladonna—Congestion to the head. Hot, red face. Eyes staring. Pupils dilated. Starting and jumping.

Camphor—(One drop in a glass of water.) When there is a coldness of the skin; cold feet and hands, pale face, with sudden difficulty of breathing. Great coldness and sleepiness.

Lachesis—Can bear no pressure on the throat or chest. Must sit up or lie on the right side. Numbness of left arm.

Cactus grand—Palpitation; fluttering. Sensation as if a tight band was around the heart.

Nux Vom—Palpitation with frequent belching. From mental emotions; from long study; after eating. Worse lying down.

Phosphorus—Great difficulty of breathing; tightness across the chest, especially in those who are growing fast.

Pulsatilla—Palpitation in young girls at the age of puberty. In persons of a sad, timid, yielding disposition. After suppression of menses.

Sepia—Tremulousness. Yellowness of the face. Yellow across the nose. Women with dark hair. During pregnancy. After suppression of menses.

Rhus tox—Palpitation worse when quiet.

Lilium tig—Sensation as if heart was squeezed and then relaxed. Great bearing down in lower part of abdomen.

---

1. Dyspepsia or Indigestion. Inflammation of the Stomach, (*Gastritis.*) and Cramp of the Stomach, (*Gastralgia.*)

By DYSPEPSIA is meant simply *indigestion*. The

symptoms are well known; heart burn, sour stomach, spitting up of food, and pain are among the more common. The sufferer becomes melancholy and morose; sleeps badly, dreams much and is miserable generally. The causes are avoidable in most cases. Eating to hastily and irregularly, swallowing the half-chewed food with liquids, over-eating, *whiskey*, *tobacco*, and *patent medicines* are the causes of nine-tenths of the cases.

INFLAMMATION OF THE STOMACH.—Is character-ized by fever; great anxiety; heat and pain in the pit of the stomach, increased by eating or drinking; vomiting and hiccough. May be caused by taking cold; irritating substance taken into the stomach; the free use of ice water, &c.

CRAMP OF THE STOMACH.—Is characterized by pains which come at intervals, with freedom from pain during intervals. Belching of wind general-ly relieves. The attacks usually follow some im-prudence in eating or drinking; or may be caused by depressing mental emotions.

TREATMENT.—In DYSPEPSIA all causes should be avoided. The quantity of food should be reduced until the stomach can dispose of it readily, and it should be taken regularly, not oftener than once in four or five hours.

Aconite—Great thirst. Vomiting. Fear of death. Hot, dry skin.

Arsenicum.—Burning pain. Thirst, drinks little at a time. Throws up whatever is taken. Rest-lessness. Very weak. From ice, ice water, ice cream, vinegar, liquors.

Belladonna—Pains compel the patient to bend backward and hold the breath. Great thirst but drinking makes it worse. Face red and hot, bloat-ed. Pupils enlarged.

Bryonia—Stitching pain worse from motion, or a

misstep. Pressing pain as of a load or stone. Thirstlessness or great thirst, drinking large quantities at a time. Constipation. Stools large and dry. In warm weather. Better when lying quiet onthe back.

Carbo veg—Great deal of belching, sour and rancid; bloatedness; disgust for meat. Burning pain down to small of back and up to shoulders. After rich food or excessive drinking.

Chamomilla—Bitter taste. Vomiting of bile or green slime. Face red and hot, or one cheek red. Very irritable.

Colocynthis—Violent, cutting, tearing pains which seem to concentrate in the pit of the stomach. Relieved by bending double and hard pressure. After vexation and indignation. Worse from potatoes.

Hydrastis—Dull ache in stomach. Weak, faint, gone sensation at pit of stomach.

Ipecac—Constant nausea with empty belchings and much saliva ; easy vomiting; diarrhea. After sour, unripe fruit, rich cake, cheese, &c.

Iris vers—Great burning distress at pit of stomach. Burning from stomach all the way to mouth. Vomiting. Diarrhea. Weakness. Sweetish vomiting.

Nux Vom—Dizziness. Headache. Constipation. Pain of a constricting character. Clawing pain extends to small of back or to the lower end of bowels, which is drawn in. Better from bending forward and rubbing the pit of the stomach. Piles, sedentary life. All worse in the morning.

Podophyllum—After eating spits up the sour food, with hot, sour belching; diarrhea. Vomits food with craving appetite afterwards. Tongue shows prints of teeth. Aching behind the eyes; white of the eye yellow. Pain when vomiting causes screams.

Phosphorus—Spits up food in mouthfuls with
out nausea. Pains in stomach worse after eating.
Water thrown up as soon as it becomes warm in
the stomach. Food scarcely swallowed comes up
again. Unnatural hunger which eating relieves
for a short time. Gastritis with heart-burn, ending with scratching in the throat.

Pulsatilla—No appetite, no thirst. Bitter taste
in the mouth. Every thing tastes bitter. Dizzy
when rising. Weight in the stomach when waking in the morning. Worse from fruits, fats, ice-
cream, pastry.

Sepia—Belchings tasting like rotten eggs. Nausea after eating or in the morning before eating;
when riding in a carriage. Aversion to meat.
Smell of food sickens. White sediment in the
urine. Yellow across the nose.

Sulphur—Faint, gone sensation at 11 A. M.
Constipation. Piles.

Lycopodium—Little food fills him up. Belching
does not relieve. Abdomen bloated, pressure upward and downward. Red sand in urine.

———

2. Vomiting.

Of itself a mere symptom; may indicate,

Aconite—When with the vomiting there is heat,
thirst, profuse sweat, and increased flow of urine.

Arsenicum—Vomiting immediately after eating
or drinking. Burning in the stomach. Nausea
at 11 A. M. and 3 P. M.

Aethusa—Vomiting of curdled milk; of frothy,

white substance. Milk is thrown up soon after taking, then drowsy.

Ipecac—In most cases the remedy. Vomiting with thirst, sweat, bad breath; always nausea; sleepy afterward; great deal of saliva.

Mercurius—Vomiting with sweet taste in throat, dizziness, heat and headache.

Phosphorus—Soon as water becomes warm it is thrown up.

Podophyllum—Vomiting of milk in infants with protrusion of the anus.

Pulsatilla—Vomiting of blood from suppressed menses.

Veratrum—Vomiting with great nausea, thirst for cold drinks, cold sweat on forehead, and prostration.

----

3. Hiccough.

A spasmodic affection of the diaphragm; may require,

Arsenicum—Hiccough at the hour when the fever ought to come.

Belladonna—Hiccough and belching come together.

Cina—Hiccough during sleep.

Cuprum—Hiccough precedes the vomiting, or attack of asthma.

Hyoscyamus—Hiccough with spasms and rumbling in abdomen.

Ipecac—Hiccough with nausea.

Nux Vom—Hiccough from over-eating, or from cold drinks.

Pulsatilla—After cold fruit; after drinking.

Veratrum—After hot drinks.

1. Diarrhea, Dysentery, Cholera, Cholera Morbus, and Cholera Infantum

are all included in one section as the indications for remedies are the same, or nearly the same in each.

DIARRHEA affects more particularly the upper portion of the bowels. The passages are generally frequent, preceded by pain, and are generally watery, of a greenish or yellowish color.

DYSENTERY affects the lower portion of the bowels. The stools at first are diarrheaic. They soon become slimy and bloody, and are accompanied with cutting, griping, drawing-in pains. CHOLERA is caused by a specific poison. It is characterized by profuse discharges from stomach and bowels looking like rice water; the nose becomes pointed, the cheeks fall in and the eye-balls are sunken; the skin wrinkles; tongue and skin become blue; there is great want of air; the breath becomes cold; cramps and spasms which contract the muscles into hard knots. This is a faint outline of that fearful scourge.

CHOLERA MORBUS differs from Cholera in not being caused by a specific poison, and in not being so violent. It is caused frequently by checked perspiration, drinking ice-water, or imprudence in eating. It usually comes on suddenly, and often in the night. Begins with vomiting and purging of bilious matter which soon changes into the rice-water appearance; great thirst; cramps of the limbs, &c.

CHOLERA INFANTUM prevails mostly among young children during the summer. It begins usually with diarrhea, followed by vomiting. The vomiting may cease for a time to come on again. There is generally fever and thirst, though drinks are not retained. The child becomes weak; loses flesh; utters now and then plaintive cries; rolls the head; squints the eyes, and falls into a stupor.

TREATMENT—Quietude and rest in the horizontal position are important aids to recovery. A flannel bandage, broad enough to cover the whole abdomen should be worn snug.-

In many cases the food must be changed, especially with children who are being brought up on cow's milk. I have seen many cases improved by using Ridge's, Horlick's, or Neaves' prepared food. Often a change from the crowded city to the pure air of the country is all that is required to recruit the little sufferer. *See section on Teething.*

Aconite—High fever. Skin hot and dry. -Thirst. Restlessness. Fear of dying. Dizzy or faint when rising up, with paleness, face flushed when lying. Everything tastes bitter except water.

Aethusa—Vomiting of white, frothy substance, or curdled milk. Lumps of curdled milk in stool. Spasms; eyes turned downward; thumbs clenched; face puffy, spotted red. Stupor.

Arsenicum—Diarrhea worse at night. After eating or drinking. After midnight. From cold food, ice-water or ice-cream. Diarrhea after extensive burns. There is great thirst for little water at a time but often. Burning in the anus and rectum. Restlessness. Vomits after eating or drinking. Great weakness, faint and trembling.

Belladonna—Congestion to· the head; head and face hot and red. Starting and jumping. Eyes red and staring, pupils dilated. Stools involuntary; of thin, green or bloody slime.

Camphor—During epidemic of cholera. Icy coldness of the whole body. Cold clammy sweat. Face pale, blue, icy cold; upper lip drawn up, exposing the teeth. Foam at the mouth; eyes sunken and fixed. Great sinking and collapse without stool. Moans and groans in a hoarse, husky voice. Burning in the stomach and throat. Tenderness at pit of stomach. There may be no stool, no vomiting, no thirst, no nausea. For this condition of collapse give one drop of the tincture of camphor on sugar every 5 minutes until relieved.

Cantharis—Stools look like the scrapings of hogs intestines. During stool there is burning at the anus. After stool, burning, biting and stinging, with urging. Violent chill as though one had water poured over him, with internal warmth. Frequent urging to urinate, with burning during and afterward.

Chamomilla—Stools smell like rotten eggs. Child is peevish, is only stilled by being carried about. Redness of the cheeks, or one cheek.

Calcarea carb—For children with large stomachs, who sweat much about the head when sleeping; who have good appetites yet grow poor. Stools are clay like, or chalky, with a sour smell; vomiting of food. Sour food or milk comes up. Worse in fat children, with the spaces in the skull unclosed. The urine smells very strong.

Cina—Worse in the day-time; after drinking. Children cry much. Reject everything offered. Pale around the nose and mouth. Picks or bores in the nose. Urine is white, milky, jelly-like. Grinding of teeth. Gurgling in the stomach when drinking. Child keeps swallowing. Gagging.

Colocynthis—Stool first watery and slimy, then bilious, lastly bloody. Cutting colic and great urging before the stool. During the stool : colic, headache, griping, nausea. Griping, cutting, squeezing pains which cause the patient to bend double and are relieved by pressure on the abdomen. Pains extend down the thighs.

Cuprum—Vomiting and purging with violent cramps all over, particularly in the stomach and limbs, with piercing screams. Face drawn out of shape. Eyes sunken with blue rings around them. Sweetish taste in mouth. Cholera.

Ipecac—Before the stool : nausea, vomiting. During stool : nausea, vomiting, coldness, paleness. Vomiting of food, mucus; of green, jelly-like mucus. Sleeps with eyes half open.

Kali bichrom—Every year at the same time. Early part of summer. Tongue coated with a thick yellow or brown fur at the root ; or tongue is dry, red, smooth with cracks.

Magnesia carb—Stools green, watery ; with green scum like that of a frog-pond. Little white masses like pieces of tallow in the green watery stool. Stools smell sour ; frothy.

Mercurius—Stools are mostly slimy and bloody. Sour-smelling night sweat, particularly on the head. Child's thighs are cold and clammy at night. Canker sores in mouth and on tongue. Great thirst yet saliva runs from mouth. Frequent urging to pass urine, pain when so doing. Before stool: violent urging, trembling, and faintness. After stool : *a can't-get-done* feeling, followed by chilliness.

Nux Vom—Before the stool : cutting about the navel ; backache as if broken. During stool : cutting ; backache. After stool : pains and urging cease. Desire for fat food. Nausea in the morning and after dinner. Worse generally in the morning. Diarrhea and constipation alternate.

Podophyllum—Stools watery, soaking in the diaper leaving a sediment like meal ; or they may be yellow or white, slimy ; or greenish, watery. Black stools in the morning. Headache alternates with diarrhea. Restless sleep with half closed eyes; moaning, grinding the teeth, rolling the head from side to side. Gagging or empty retching. Anus protrudes. The watery stools are generally painless and come with a gush. Diarrhea worse early in the morning. Flesh cool with sweat on the head. Stool passed while being washed.

Phosphorus—Stools are very profuse and watery, pouring out as from a hydrant. Bloody stools with small white particles like little pieces of tallow. Chronic painless diarrhea of undigested food with much thirst for water during the night. Vomiting of what has been drunk as soon as it becomes warm in the stomach. Sleepy in the daytime and after meals.

Pulsatilla—Stools during the night mostly. Aversion to fat food, meat, bread, milk. No thirst. Bitter taste in the mouth. During stool: shaking chill; pain in small of back.

Rheum—Thin, watery, sour-smelling stool. Frothy, of a pea-green color. Cool sweat around the nose and mouth. Wants various kinds of food which become repugnant as soon as a little is eaten. Sour smell of the child in spite of the utmost cleanliness.

Rhus tox—Stools look like water in which meat has been washed. Tearing pains down the limbs. Restless sleep, dreaming of hard work and difficulty. Thirst for cold milk.

Sanguinaria—During diarrhea much wind escapes upward and downward. Diarrhea after cold in the head, pain in the chest or cough.

Sepia—Diarrhea worse after boiled milk, or meat. Stools are jelly-like. Frequent waking at night. Wakes at 3 in the morning, cannot get asleep again.

Sulphur—Heat on top of the head. Burning in

the soles of the feet. Gets faint and hungry in the forenoon at 10 or 11. Sleepiness in the afternoon, after sunset. Ears red. Very sore, red, and tender around the anus. Does not like to be washed. Tongue with red tip and sides.

Veratrum—Most important in cholera and cholera morbus. Vomiting and purging coming on suddenly in the night. Discharges gushing, profuse, like rice-water, with cramps. Cold sweat on forehead. Face looks deathly. Vomiting and cold sweat during stool. Great weakness. Extreme thirst for cold water or sour drinks.

---

2. Constipation.

This trouble prevails mostly among females, and men of sedentary habits. It is due in a great measure to neglecting and delaying the regular daily movement of the bowels. It is often the result of a pernicious habit that many have of resorting to pills, salts, castor oil, rheubarb, &c. ; the ultimate result of which is to weaken and destroy the tone and vigor of the intestinal canal.

TREATMENT—First of all, break up the habit of delay. Every morning, make an effort for an evacuation. A glass of tepid water drank early in the morning is sometimes beneficial. A glass or two of slippery-elm mucilage is perfectly harmless and has often done good, taking it morning and night. Smartly striking or percussing the entire abdomen, from the ribs downward, morning and night is recommended in the "movement cure." Fruits and vegetables should form a good part of the diet. Right here I wish to condemn the practice of eating coarse bran, either in graham bread or otherwise. The coarse, hard shell of the grain is indigestible and acts as an irritant to the stomach and bowels.

Bryonia—Stools hard, dry, as if burnt; of large size. Troubled with rheumatism. Irritable, easily angered.

Graphites—Stools large and knotty; lumps united by stringy mucus. Sometimes much slime is expelled with the stool. Sharp cutting pain during stool, as though the parts were torn, followed by constriction and aching for hours. Eruptions on the body which emit a sticky fluid.

Hydrastis—Faint, sinking, gone feeling at the pit of the stomach.

Hepar—Constipation with eruption in bend of elbow or knee.

Lycopodium—Urging without avail; the rectum seems closed. Wind accumulates in the bowels. Red sand in the urine. Little food fills him up.

Nux Vom—Constant urging without success. Stools large and hard. Rush of blood to the head. Worse from high living and sedentary habits.

Opium—Constipation of corpulent, good humored women and children. Stools in hard, round, black balls. No urging for stool until there is a large accumulation. Dull and sleepy.

Phosphorus—Stools long, slender, hard, dry, and difficult to expel. Painful cramps in rectum after stool. Tall, thin persons, with dark hair and eyes.

Plumbum—Stool in hard, small balls, passed with urging and great pain as if anus was constricted. Abdomen drawn in at the navel.

Podophyllum—Constipation with protrusion of the anus. Back weak and sore after doing a washing. Flashes of heat up the back.

Sepia—Hard, knotty, insufficient stool, mixed with or covered by slime. Urging to stool, only wind and mucus pass; feeling as if a lump remained in the rectum.

Silicea—After long straining the partly expelled stool slips back again.

Sulphur—Shooting, cutting pains from the anus upward after stool. Beating pain at the anus. Soreness and stinging at the anus prevents lying or sitting. Heat in top of head. Faint and hungry at 10 or 11 A. M.

See also section on piles.

——

3. Piles, (*Hemorrhoids.*)

This disease consists of an enlargement of the veins at the anus. These form tumors from the size of a pea to that of a chestnut. When these tumors are on the outside of the anus, they are of a dark blue appearance, and are called *external piles*; when on the inside they are called *internal piles*; when, during an evacuation they bleed, they are called *bleeding piles;* when they do not bleed they are called *blind piles.* Among other causes of piles are the avoidable ones of liquors, tobacco, and the habit of taking *physic* for every trifling ailment.

TREATMENT—Keep the bowels regular. Avoid straining at stool. Should the tumors descend during stool replace them gently as possible. This can be more easily done by lying on the face with the hips elevated. For an acute attack, with great soreness and heat, small pieces of ice slipped into the rectum, will sometimes afford relief. Lie down as much as possible.

Hamamelis—May be used locally in any case. One part of tincture to twenty parts of water.

Aesculus hip—Soreness, burning, itching, and fullness of the anus. Rectum feels as if full of small sticks. Seldom bleed. Aching, lameness or shooting in the back. Piles worse when standing, better when walking. Throbbing in lower part of body. Stool; first part hard and black, the last part softer and as white as milk.

Aloes—Piles protrude like grapes; hot and sore; relieved by cold water. When he passes urine feels as if something would escape from the bowels at the same time.

Hamamelis—Piles bleed profusely, with burning, soreness and weight at the anus. Back pains as if it would break. *Hamamelis* may be applied locally with a syringe or sponge, in any, case, diluting one part of the tincture with twenty parts of water.

Lachesis—At the change of life. Sharp, cutting pain in the piles whenever sneezing or coughing.

Mercurius—Bleeding from piles whenever passing urine.

Nux Vom—Alternate constipation and diarrhea. In persons who have had much drugging, or have used much tea, coffee, liquors or tobacco; or high living.

Collinsona—Piles with constipation. Sensation as if sticks, sand or gravel had lodged in the rectum.

Podophyllum—Piles with protrusion of the lower part of the bowel; with diarrhea worse mornings. Discharge of thick transparent mucus, or mixed with blood.

Sulphur—Frequent desire for stool without success. Cutting pain from anus upward after stools, and between stools. Heat in top of head. Burning in soles at night. Faint feeling at 10 or 11 A. M.

See also Constipation.

4. Colic, (*Enteralgia.*)

Cramping, griping, twisting pains that come in paroxysms with more or less freedom from pain during the intervals.

TREATMENT—Warm applications and pressure usually afford relief.

Aconite—After taking cold. Fears he will die.

Arsenicum—Pains worse at night. Burning in stomach and abdomen. Drinks little and often. After the use of ice-water, ice-cream, bad sausage or cheese.

Belladonna—Pain as though the intestines were seized with the nails and clawed together. Pains come suddenly and •leave suddenly. Congestion to the head.

Cocculus—Wind colic at midnight. Sensation as if sharp stones were rubbing together on every movement. Belching relieves.

Colocynthis—Pains around the navel; or pains from different parts concentrate in the pit of the stomach. Has to bend double and press against the abdomen. Brought on by vexation or anger.

Cuprum—Colic with abdomen drawn in.

Lycopodium—Colic on right side extending to the bladder. Red sand in the urine. Belching does not relieve; passing wind downward relieves.

Nux Vom—Wind colic with pressure upward on the chest causing difficult breathing, and downward causing urging to stool and to urinate. Colic with water brash; worse after coffee, liquors or over-eating; worse before breakfast and after meals.

Pulsatilla—Wind colic, has to bend forward. Pressure in abdomen and back as from a stone; lower limbs go to sleep when sitting. From ices, fruits, pastry, getting the feet wet.

Sulphur—Intestines feel as if strung in knots; worse from bending forward.

————

5. Inflammation of the Bowels, (*Enteritis, Peretonitis.*)

Usually caused by imprudence in eating, taking cold, or external injuries. There are sharp cutting pains or tenderness through the abdomen, hiccough, vomiting, diarrhea or constipation, bloating and fever.

TREATMENT—*Aconite*—Skin dry and hot. Thirst. Restlessness. Mouth and tongue dry. Bitter taste. Fear of dying. Fixes the time. Urine scanty, red and hot. Feet cold.

Arsenicum—Great loss of strength. Cold, clammy perspiration. Thirst, drinks little and often. Burning in the bowels. Worse about midnight. Restless.

Belladonna—With congestion to the head. Throbbing in sides of neck. Cant bear noise or light, nor least jar of the bed. Pains come and go suddenly. Cannot bear the slighest touch. Delirum.

Bryonia—Stitching pains through the bowels, worse from the slightest motion. Great thirst, drinks much at a time. Patient lies perfectly still, don't want to move.

Cantharis—Abdomen burning hot, bloated. Burning pain in anus after stool. Frequent painful desire to urinate with smarting and burning afterwards.

Rhus tox—Restlessness. Changes position often although it causes pain. Tongue red at tip.

6. Inflammation of the Liver, (*Hepatitis.*)

Usually due to taking cold, or to external injuries. There is pain in the right side of abdomen just below the ribs; pain in the right shoulder; swelling of the liver; yellowness of the skin; chills and fever.

TREATMENT—*Aconite*—High fever. Hot, dry skin. Quick pulse. Thirst. Great anxiety. In the beginning.

Bryonia—Stitching pains, worse from least motion. Great thirst for large draughts.

Leptandria—Black stools. Aching in liver. Yellow or dark coating on tongue.

Mercurius—Cannot lie on right side. Thirst, yet mouth and tongue are moist. Sweats at night without relief.

Podophyllum—Principal remedy. Heat in region of liver. Wants to rub the side with the hand. Skin and eyes yellow. Diarrhea, or stools may be pale, hard and dry. Vomiting of bile. Fullness and pain in right side. .

---

7. Worms, (*Helminthes.*)

1. PIN WORMS or SEAT WORMS are from 1-4 to 1-2 inch in length, and inhabit the lower part of the rectum. They are found more often in children than grown people. They cause great tickling and irritation in the anus and genital organs. They crawl into the vagina in females, and under the fore-skin in males. They may produce a discharge from those parts; wetting the bed at night; fits; or the habit of masturbation.

2. The ROUND or STOMACH WORM is from six to twelve inches long, pointed at both ends, large as

a goose-quill, and inhabits the small intestine and stomach. They sometimes cause gripings, slimy stools, irregular appetite, itching of the nose, dilated pupils, squinting, starting, grating the teeth, and perhaps convulsions.

3. TAPE-WORM. This worm also inhabits the small intestine. Its head is about as large as the head of a pin; neck, 1-2 inch long; body may be 10 feet or over in length, and consists, of links or segments which grow broader from the head to the tail. The larger segments fall off from time to time, and are discharged with the stool. Sometimes the following symptoms may be complained of: Pains in the stomach; nausea; vomiting; great hunger, even to fainting. There may be dizziness; getting dark before the eyes; palpitation; chorea; epilepsy.

TREATMENT—In the case of pin-worms, injections into the rectum and vagina of a solution of salt in water, or 2 or 3 drops of carbolic acid in 1-2 pint of water. One-fourth to one-half pint of solution should be used at a time, and repeated three times a week. For stomach worms I know of no better way to expel them than to take 3 grains of *Santonine*, rub it with 9 grains of white sugar, divide into 12 parts, giving *one part* for each year of the child's age, at a dose. It may given once in 3 hours for 12 hours, then follow with some cathartic. This course may be repeated. The ordinary "Worm Lozenge" contains Santonine, and may be used. Or the "*Fluid extract of Spigelia and Senna*" in doses from 1-2 to 2 teaspoonsful. For tape worm, the inner part of *squash seeds* or *pumpkin seeds* eaten freely at night and in the morning, and followed with a cathartic has often succeeded. In one instance the use of pumpkin seeds resulted in the expulsion of a host of pin worms. *Kousso* is also used. For worm troubles in children, consult the sections on *Convulsions*, *Teething*, *Diarrhea*, *&c.*

---

1. Jaundice, *(Icterus.)* Biliousness.

The most characteristic symptom of JAUNDICE is the yellow discoloration of the skin. With that are apt to be present : itching of the skin; slowness of the pulse; yellow sight; night-blindness; urine that will stain yellow; pale, whitish stools; headache, dizziness &c. Jaundice occurs in the latter months of pregnancy, in new-born children, and sometimes after violent emotions, especially anger.

The symptoms that usually constitute the condition called BILIOUSNESS are: sallow face; yellow coated tongue; loss of appetite; headache over the eyes; drowsiness; languor; constipation; depression of spirits, &c.

TREATMENT—Abstain from all physic, bitters &c. A light diet with plenty of fruit. Bathe and brush the skin thoroughly every day or two.

Aconite—Tongue coated white or yellowish. Bitter taste in mouth; everything tastes bitter except water. Dizzy when raising the head. Desire for beer, or bitter drinks. Stitches in region of liver. Constipation, with clay-colored stools. Skin yellow.

Eupatorium perf—Dizzy in the morning. Heaviness in back of head. Heat on top of head. Tongue yellow or white; taste bitter. Vomiting of bile. Soreness in region of liver, can't bear the clothing tight. Aching in back and limbs. Sleepy at noon. Chilly in morning.

Mercurius—Head feels as if in a vice, with nausea; worse at night in bed. Sour smelling sweat on head. Yellowness of white of eye. Tongue slimy, or coated yellow; flabby, showing prints of teeth. Face yellow. Hungry, but relishes nothing. Sore in region of liver; cannot lie on right side. Sweats at night. Feels worse at night. Skin dirty yellow, rough and dry. Constipation, stool tough, or crumbling. Constant urging, worse at night. Drowsy, irritable, taciturn.

Nux Vom—Disinclined to work, lassitude and weakness in the morning. Inclined to find fault and scold. Dizzy in morning. Headache. Lower part of eye-balls yellow, or more so than upper part. Bitter, sour, or putrid taste in mouth. Tongue coated white or yellow. Sour or bitter belchings. Cannot bear clothing tight about the waist. Yellow skin. Faintish spells. Constipation. Piles. Drowsy after meals. Worse mornings; in persons who get but little out-door exercise.

Leptandria—Jaundice with clay-colored stools Nausea and faintness on rising. Tongue coated yellow or black through the centre. Burning distress in liver. Pain in right shoulder. Weak feeling in bowels.

Podophyllum—Depression of spirits. Thinks he will die or be very ill. Dizzy, with full feeling over the eyes. Dull headache. Darts of pain through forehead. Tongue yellow, or white and moist, shows prints of teeth. Everything tastes sour. Fullness from small quantity of food. Hot, sour belchings. Eye-balls yellow; aching behind the eyes; whitish stools. Early morning stool. Rumbling in bowels with drowsiness, more in forenoon. Protrusion of anus. Piles. Pain between shoulders, in morning.

See also Hepatitis, Dyspepsia, Headache.

1. Inflammation of the Kidneys, (*Nephritis.*)

Generally caused by exposure to cold, external injuries, or drugs that produce irritation of the urinary organs. It commences with a chill usually, fever, nausea, vomiting, pain and heat in the small of the back, extending to the bladder. Frequent desire to urinate, with but little urine, of red or brown color. There is often a puffiness of the eyelids, face and extremities.

TREATMENT—*Aconite*—High fever. Restlessness. Scanty, high colored urine. From exposure to cold.

Camphor—(Five drops of tincture to a glass of water.) The best remedy when the difficulty has been brought on by the use of blister plasters, cubebs, copavia or other diuretics.

Cantharis—In most cases. Urine flows drop by drop, with burning and pain. Feels as if he could not get through.

BRIGHTS DISEASE may be defined in short, as a chronic inflammation of the kidneys. Pain in the small of the back; bloating of various parts of the body; paleness of the skin, and inside of the mouth, and of the tongue, and an unaccountable weakness, should warn the individual to lose no time in placing his case in the hands of a competent Homeopathic Physician. We would say the same in regard to

DIABETES, a disease characterized by great thirst and the passing of an excessive quantity of urine.

---

2. Inflammation of the Bladder, (*Cystitis.*)

Caused by cold, irritating drugs, injuries, retention of urine, &c. Symptoms are pain in region of the bladder, worse from pressure and motion; frequent and painful passing of urine, which is expelled drop by drop with straining, and may be bloody; fever. Chronic cases should be treated by a Physician.

TREATMENT—*Aconite*—In the beginning of most cases. High fever. Urine hot and red or turbid.

Arsenicum—Burning pain when beginning to pass urine. Cold sweat. Face, hands and feet cold. Worse midnight and after. Drinks little and often.

Belladonna—When *Aconite* does not relieve. Region of bladder very sensitive to touch and jar. Delirious.

Lycopodium—Before passing water child screams with pain. Red sand on diaper.

Camphor—See preceding section.

Cantharis—The main remedy in most cases. Is usually given alone or in alternation with other remedies.

———

3. Dribbling or Involuntary Urination, (*Enuresis.*)
Wetting the Bed, (*Enuresis Nocturna.*)

TREATMENT—When children wet the bed at night they should take but little fluid or food of a juicy nature in the afternoon. The bladder should be emptied the last thing before going to bed. Suspect the presence of worms. Dribbling with the youth or adult would lead us to suspect a stricture or gravel; with the aged it might indicate paralysis of the bladder; conditions that would require professional aid.

Belladonna—Wets the bed. Restless. Starts in sleep. In florid children.

Arsenicum—In pale, feeble children, who are cold all the time. They drink but little at a time. Sometimes look puffy under the eyelids.

Cina—Urine milky. Worm symptoms.

Pulsatilla—Involuntary urine when coughing, sneezing, passing wind, or during night. In little girls. Menses suppressed.

Sepia—Urine escapes during the first sleep. In females with suppressed menses, or who are troubled with leucorrhea.

Equisetum and *Eupatorium purp* are recommended and may be tried.

---

4. Retention of Urine. Difficult Urination, *(Dysuria.)*

May be due to stone in the bladder, strictures, paralysis, &c. Sometimes children push foreign bodies up in the passage.

TREATMENT—Mechanical obstructions will require to be removed. Warm water injected into the rectum is often of service. All so called *diuretics*, or *forcing* medicines are worse than useless.

Consult the indications in sections 1 *and* 2 ; *also,*

Lycopodium—Urging, must wait a long time before it will start. Red sand. Child screams with pain before urinating.

Lithium carb—Pressure about the heart, relieved after passing urine.

---

5. Inflammation of the Testicles. *(Orchitis.)*

This is caused usually by external injuries, or may result from *mumps*.

TREATMENT—Quietude. If from a bruise or blow, *Arnica-tinct* may be diluted with water (1 part to 10) and applied externally.

Aconite—If from taking cold, or if there be fever.

Arnica—(1 drop of tincture to a glass of water) if from a blow or bruise.

Belladonna—Swelling bright red. Pains worse from least jar or motion. Head symptoms.

Rhus tox—Dark red; from a strain; working in the wet; from mumps.

Pulsatilla—From a cold; after mumps.

SPERMATORRHEA—An involuntary discharge of semen. Caused by secret bad habits; sexual excesses; eruptions on the genital organs; worms; costiveness &c. A disease by no means incurable, yet *woe* to the unfortunate victim who passes through a course of Allopathic treatment for it.

———

1. Rheumatism. Sciatica. Lumbago.

RHEUMATISM affects the joints and muscles principally. When acute it is usually attended with fever, pain, tenderness and swelling of the affected part. When chronic there may be little or no fever or swelling; the joints are apt however to become enlarged, misshapen and stiffened.

LUMBAGO is a rheumatic affection of the muscles of the back.

SCIATICA is characterized by a pain extending down the posterior part of the thigh from the hip to the knee, and sometimes below. Often the pain is paroxysmal, with a tingling and numbness of the limb.

TREATMENT—Wrapping the painful limb or joint in cotton batting will often give considerable relief.

Aconite—Fever, thirst, restlessness. Urine scanty, red and hot. Swelling of the joints. Stiff neck, pains down the neck to right shoulder.

Belladonna—Pain runs from joint along the limb like electric shocks. Worse at night and from least touch or jar. Red, shining swelling of joints. Headache. Beating in the neck. Backache as if broken.

Bryonia—Stitching pains from least motion. Dryness in mouth with or without thirst. Dry stool, as if burnt. Dizzy and sick when rising up. Sour, greasy sweat. Right side of neck stiff and painful.

Chamomilla—Drawing, tearing pains from left hip down to heel and sole of foot. Burning in soles. Must keep quiet.

Colocynthis—Crampy pain in the hip as if the parts were squeezed in a vice ; lies upon the affected side with the knee bent up. Cramp in calves of legs. Pains in the thumbs.

Kalmia lat—Rheumatic pains generally go from the upper to the lower parts. Violent pain in upper part of back through the shoulder-blades. Pains shift from joint to joint. Pains worse on motion and earlier part of night. Pains suddenly leave the limbs and go to the heart. Muscles of neck sore to touch and movement. Tearing pains from hip down to foot.

Ledum—Pains go from below upward ; joints pale, swollen, tense and hot : worse from warmth of bed, from motion, and in evening before 12 P. M.

Mercurius—Tearing, stinging pains, worse at night ; with profuse sweat which gives no relief.

Phytolacca—Both shoulder blades ache. Shooting pains in shoulder-joint, with stiffness and inability to raise the arm. Aching and tenderness on top of shoulder. Pains shooting down the outside

of hip and thigh.   Sharp, cutting pain, drawing the
leg up; cannot touch the floor.   Pains from hip to
knee, worse on outside of thigh.   Feet puffed, soles
burn, pains in ankles and on the top of foot.   Pain
in heel.   Urine stains the vessel red.   Stiff neck.
Backache with sore throat.

Pulsatilla—Pains shift from place to place; worse
at night; from warmth,   Redness and swelling of
joints, with stinging pains.   Aching in calves,
which are swollen.   Feet swollen, top and sole.

Rhus tox—Tearing, shooting pains; dull, aching
pains; all worse at night, and in damp, cold weather; better from rubbing, heat, and when warmed
up by exercise.   Numbness and crawling sensation
in affected parts.   Pains worse on beginning to
move, better from continued motion; worse after
midnight.   Stiff neck.   Pain in small of back, better lying upon something hard.

Nux Vom--Pains dart from toes to hip, or from hip
to hollow of knee; worse at stool, from motion, lifting
and at night.   Cramps in calves and soles at night;
must stretch the foot.   Pain in small of back, worse
3 to 4 A. M.; must raise up to turn over in bed.

Podophyllum—Stiff neck, with soreness of muscles.   Pain between shoulders in the morning; under right shoulder blade.   Backache after washing.

Sanguinaria—Rheumatism in right arm and
shoulder; cannot raise the arm; worse at night,
or on turning in bed.   Pain in left hip; inside of
right thigh.   Pains run down the limb.   Leg and
foot swollen with burning pain; limb cold outside;
worse until midnight.   On touching the painful
part pain vanishes to appear at some other part.
Pain in the back from lifting.   Pain in neck, shoulders and arms; worse at night.

Sepia--When stooping, sudden pain in the back,
relieved by pressing back against something hard.

1. Dropsy, [*Anasarca.*)

This disease consists in an effusion of watery fluid into the loose tissue beneath the skin, or into the cavities of the body. When the effusion is under the skin, the disease is characterized by swelling of the parts, with pitting when these are pressed upon. When the effusion is into some of the cavities the symptoms will vary according to locality. If into the cavities of the brain, there will be spasms, convulsions, stupor, &c. If into the pleural cavity, there will be difficulty of breathing, cough &c. DROPSY is never a primary disease. It nearly always depends upon and follows some disease of the lungs, heart, liver or kidneys. It frequently exists with or follows scarlet fever.

TREATMENT in serious cases should be undertaken by an intelligent Physician only.

Apis—May be used when the urine is dark, like coffee grounds ; scanty. Stinging, burning pain in various places. Bag-like swelling under the eyes.

Arsenicum—Pale or greenish complexion. Feels faint from slightest motion. Thirst, drinking often but only a small quantity at a time. Suffocative spells at night, must jump out of bed. Skin cool, with burning heat inside.

Digitalis—Dropsy with difficult urination. Pulse slow when at rest, but becomes rapid from the least motion. Pulse intermits every third, fifth or seventh beat.

Helleborus—Especially after scarlet fever. Urine scanty, dark, with dark specks floating through it. Jelly-like stools. Drowsy.

China—Dropsy after loss of blood ; or in nursing women.

Lachesis—Lower limbs bloat, first left, then right. Urine almost black.

Apocynum cann—Dropsy with great thirst, but water causes pain or is vomited.

---

1. Measles, *(Morbilla.)*

Contagious; occuring, as a rule, but once in a life; confined mostly to children. In from eleven to fourteen days after exposure it shows itself as an ordinary cold. There is chilliness, feverishness, sneezing, watery eyes, cough &c. This continues for 3 or 4 days, when the fever increases, and occasionally convulsions occur. An eruption now appears first upon the face and spreads downward over the body, and a peculiar smell scents the air around the patient. On removing the finger after having made pressure for a moment, the redness reappears from the centre toward the outer edge of the white spot.

TREATMENT—*See general Directions.*

Aconite—In the outset if there is a hot, dry skin, fever, &c.

Arsenicum—Burning in the skin; quick small pulse. Great thirst, drinking but little at a time. Face bloated. Vomiting and diarrhea. Great weakness. Worse about midnight.

Belladonna—In the beginning with a hot moist skin. Drowsy sleep, or drowsy but cannot sleep. Pupils enlarged. Congestion to the head.

Euphrasia—Hot, burning tears from the eyes; profuse flow from the nose not burning. Cough during the day.

Mercurius—Canker sores in mouth. Slimy diarrhea.

Phosphorus—Dry, tight cough, worse from evening until midnight. Painless diarrhea.

'*Pulsatilla*—Nightly diarrhea. Thick yellow discharge from nose. Loose cough with thick yellow mucus. Dry mouth without thirst.

---

2. Scarlet Fever, (*Scarlatina.*)

A highly contagious disease which affects principally the *skin*, the *throat* and the *kidneys*. All ages except infancy are liable to contract the disease, scarcely ever the second time however. In from 8 to 12 days after exposure the patient is seized with chills, fever, nausea, vomiting, headache and prostration. Also soreness and dryness of the throat with burning and pain when swallowing, are complained of. After an interval of a day or two an eruption shows itself on the neck at first, spreading within 24 or 36 hours over the whole body. A white spot from pressure with the finger becomes red again from the outer edge to the centre, unlike measles in this respect. During this period the symptoms become more severe and convulsions often appear. This stage may continue for 4 or 5 days when we will find white scales peeling off the neck, as occurs to the entire surface usually. In favorable cases the symptoms now begin to abate, and in 3 or 4 weeks from the outset of the disease perfect recovery may take place. It is often accompanied or followed by deafness or by dropsy, the result of inflammation of the ears or kidneys.

TREATMENT—The intense itching may be allayed by greasing with the fat of ham or bacon. Those liable to take the disease should be excluded from the room.

See General Directions.

Belladonna—Delirum. Drowsy but cannot sleep. Starts in sleep. Involuntary moving of the hands to the head. Pupils dilated; eyes red and staring. Skin burning hot. Bends the head backward. Throat sore, swollen inside and outside; can hardly swallow.

Arum triph—Lips and mouth raw, sore and bleeding. Corners of lips cracked and bleed.

Rhus tox—Eruption looks dark. Great flow of tears. Restlessness. Nose bleed at night. Rheumatic pains worse when perfectly quiet.

Ammon carb—Eruptions more on upper half of body. Glands under right ear and on right side of neck swollen.

---

3. Small Pox, (*Variola.*)

No age or sex is exempt from this extremely contagious disease, yet it rarely, if ever, attacks the same person twice. It sets in, nine days after exposure, with a chill, or chilly sensations, followed by permanent heat. Other unpleasant symptoms occur, of which the *dreadful backache* is the most characteristic of the disease. On the third day little red spots or pimples appear first on the face, spreading thence over the body. These pimples develop first into vesicles and then into pustules filled with a milky, purulent fluid. Each fully developed vesicle or pustule has a little depression at its summit. Many other organs than the skin may be affected during the course of the disease. During the last stage the pustules burst; their contents dry into hard scales which finally drop off, leaving in many instances scars which remain for life. *Varioloid* is simply a mild form of small-pox.

TREATMENT—The patient should be vaccinated at once, although he may have contracted small-pox, as it no doubt modifies the disease. He should be removed from the neighborhood of others if possible, and all communication with him by those who are unprotected, prohibited. All bedding, garments &c. used should be destroyed.

Arsenicum—Great prostration, yet very restless. Great thirst, drinking little but often. Burning heat. Vomiting. Worse about midnight.

Belladonna—First stage.

Mercurius—Great flow of saliva. Slimy stools.

Phosphorus—Contents of pustules bloody. Hard, tight cough. Hemorrhage from the lungs.

Tart em—Nausea. Vomiting. Drowsy. Warm sweat on forehead.

---

4. Chicken Pox, (*Varicella.*)

Consists at first in little red spots like a flea-bite which soon develop into little vesicles and pustules, the latter of which may leave a scar. They usually appear in crops for several days. It rarely needs treatment. If needed, *Aconite*, *Belladonna*, *Mercurius*, or *Tart em* may be given.

---

5. Erythema.

ERYTHEMA is characterized by a diffused redness of the skin, which under pressure of the finger leaves a yellow spot which becomes red again at once. It is caused by heat, exposure to the hot rays of the sun, irritating substances as mustard

&c. It is the ordinary chafing so common to babies and fleshy people in hot weather. It sometimes comes upon the lower extremities of young persons, and shows itself as lumps upon the reddened skin, resembling bruises very much.

TREATMENT—When it occurs between the thighs of an infant, the child should be washed and wiped dry immediately after each passage of urine or stool. Very finely powdered starch is sometimes sprinkled over the raw surface. A solution of borax is recommended as an application.

Cantharis—If the urine seems to scald, or smells very strongly.

Lycopodium—Red stain upon the diaper.

Graphites, Petroleum--If eruption comes behind the ears.

---

6. Salt Rheum, (*Eczema.*) Shingles, (*Herpes zoster.*) Ring-worm, (*Herpes circinatus.*)

All consist of vesicles upon an inflamed surface. These vesicles may dry up forming scales, or they may burst and their contents dry into crusts and scabs, with chaps and cracks.

SALT RHEUM and RING WORMS may show themselves upon any part of the body.

SHINGLES however, usually develop themselves upon the chest or neck, extending in a belt half way around the body, and are preceded by rheumatic pains for a few days.

TREATMENT—*Arsenicum*—Scaly eruptions; nightly burning or itching. Better from external warmth. Restlessness.

Graphites—Worse on left side. Cracks. Chaps Sticky discharge.

Hepar—Worse mornings and on right side. Eruption spreads by means of new pimples appearing just beyond the old parts.

Mercurius—Yellow crusts. Shingles around the abdomen. Itching worse in damp weather; at night in bed.

Rhus tox—Itching worse on hairy parts; burning after scratching. Eruptions alternate with pains in chest and dysenteric stools. Shingles.

---

7. Hives. Nettle Rash, (*Urticaria.*)

This is characterized by smooth patches or blotches raised a little above the surrounding skin. These patches may be redder or paler than the skin about them. Among the known causes are certain kinds of food, as strawberries, clams, mushrooms &c.; contact with nettles and some kinds of caterpillars; the bites or stings of insects, &c.

TREATMENT—A solution of *Chloral* externally is said to relieve the intense itching and burning. This is not strange to the Homeopath, for *chloral* has often produced an *Urticaria* when taken internally.

Arsenicum—Burning; chills and fever.

Belladonna—During menstruation.

Bryonia—Rheumatic pains worse from motion.

Dulcamara—Burning after scratching. Worse from warmth, better from cold. Nettle rash alternates with asthma. .

Hepar—Nettle rash during intermittent fever. During chill.

Kali carb—During menstruation.

Pulsatilla—With diarrhea; from pastry; from pork; from delayed or scanty menses.

Rhus tox—During rheumatism; during intermittent fever; from getting wet. Worse in cold air.

Sanguinaria—Nettle rash before nausea.

Sepia—Breaks out in cold, disappears in warm air; especially on face, arms and chest.

---

8. Boils, *(Furunculus.)* Carbuncle. Felon. Run-Around, *(Panaritium.)*

The causes of these well known complaints are but little understood.

TREATMENT—External warmth and moisture in the shape of poultices seem to hasten suppuration, and often afford some relief from pain.

Apis—Stinging pains.

Arsenicum—Burning pain. Great weakness, restlessness and thirst. Worse about midnight; better from external warmth.

Belladonna—Bright redness. Throbbing pain. Drowsy but cannot sleep.

Lachesis—Dark blue, purplish appearance. Many small boils or pimples around the larger one.

Rhus tox—Boils with rheumatic pains. Feels best when moving about.

Sulphur—Boils about the fleshy part of the hips and thighs.

---

1. Typhoid Fever.

Caused by putrefying animal substance; by foul air from sewers or drains; by drinking water poisoned by soakage from drains, privies, &c. The patient can hardly fix the time of the attack. He wiil feel badly for a week or more ; then he may have a chill or chilly sensations followed by heat. From this point the heat of the body increases ; there is well marked fever, dry skin, thirst, quick pulse. The patient is weak, has no appetite, the tongue soft and flabby, showing the prints of the teeth. There is headache, dizziness, and ringing in the ears ; sleep is restless and troubled with dreams, and he talks and calls in his sleep. The bowels, which at first are costive, become loose. The face is flushed when lying down and often nose-bleed. Later there is delirium, with muttering and picking the bed-clothes ; the tongue becomes red, dry, glistening, perhaps cracked and bloody, or a brown or black strip through the centre. The abdomen becomes bloated and tender, and there is gurgling and rattling in the bowels. The patient may become more and more stupid, and he may sink into a state of complete stupor ; or he may be wild and agitated, constantly throwing off the covering and trying to get out of bed,with the delusions rapidly changing. The teeth and gums become coved with sordes, and the nostrils are blackened as of soot. The abdomen is like a drum, and the patient constantly slips down in the bed ; urine and stool are passed without his control or knowledge. When recovery takes place this condition begins to mend the latter part of the third week, the symptoms departing as gradually as they came. The three organs that suffer principally are the brain, the lungs, and the bowels ; and the symptoms will vary as one or the other organ is mostly affected. Patients are apt to lose their hair during the attack, which is soon renewed however. Profuse bleeding from the nose, bowels or womb during the later progress of the disease, is a bad omen.

TREATMENT—Study the *General Directions.* Don't

try to physic the patient in the beginning, nor to check the diarrhea later on. Let good fresh milk, given regularly, be the diet mostly. During convalescence the utmost care should be used against fatigue, over-eating, and taking cold.

Arsenicum—Great and sudden prostration. Great restlessness, constantly moving arms and legs, whilst the trunk is still on account of weakness. Picking the bed-clothes. Mild delirium. Features sunken. Lips and teeth covered with a brown or blackish slime. Tongue red, dry and cracked; stiff like a piece of wood; black. Great thirst, but drinks little at a time. Involuntary stools which smell very foul. Urine retained or passed involuntary. All symptoms worse at midnight or soon after. *Arsenicum* is useful in the majority of cases.

Belladonna—In the early stage when the brain is affected. Drowsy but cannot sleep, or frequent starting during sleep. Violent delirium; tries to bite, strike or spit on those about him. Eyes glistening, staring; pupils dilated. Throbbing in arteries of the neck. Sore throat.

Bryonia—Delirium at night of the affairs of the day or business matters. Dizzy and faint when raised up. Bitter taste in the mouth, water tastes bitter. Tongue coated thick, yellow or white. Thirst, drinking much at a time, but not very often. Cough with stitching pains through the chest. Wants to lie quiet; pain in limbs when moved. Constipation, sour sweat. Patient cross, easily offended.

Lachesis—All symptoms worse after sleep. Stupor. Muttering. Tongue red, dry or black; cracked on tip and bleeds; tongue trembles or catches on lower teeth so he cannot put it out.

Opium—Complete stupor. Can hardly be roused.

Lies speechless with open staring eyes. Pupils contracted. Face dark red and bloated. Breathing snoring, slow. Stools retained or involuntary. Urine retained. •

Phosphorus—Dry, hard cough; or loose, with clear, thick yellowish or bloody mucus Cough worse from evening until midnight. Painless diarrhea. Stools involuntary, anus seems to remain open.

Rhus tox—Answers correctly, but slowly. Nosebleed after midnight. Tongue red at tip in shape of triangle. Diarrhea worse at night. Stools pass during sleep. Limbs pain him more when keeping them still. Constant moving and changing position.

---

2. Gastric Fever.

Simply a mild form of TYPHOID FEVER, for which study remedies.

---

3. Bilious Fever or Remittent Fever.

Usually begins with a chill followed by fever, with nausea, vomiting, &c.; yellowness of skin; blisters upon the lips; whitish stools; headache; dizziness; pain in limbs; ringing in ears; weakness. There is an increase of fever every day or two, but no complete intermission as in FEVER AND AGUE.

For remedies see *following section*.

4. Fever and Ague, (*Intermittent Fever.*)

This disease is supposed to be due to a poison developed in swamps, in regions where new soil is first turned over, or where ditches or canals are dug; where the surface may be dry, yet the ground underneath full of water. This poison is called malaria or miasma. The onset of the disease is usually preceded by a general bad feeling for a few days; then occurs a paroxysm, consisting of chill, fever, and sweat; this is followed usually by an almost complete abatement of the symptoms, so that the patient is quite comfortable. The paroxysms may occur daily, or every two or three days. The different stages are not complete in every case, the chill, heat, or sweat being absent. Frequently great thirst, pain in the limbs, and headache accompany the paroxysms.

TREATMENT—To be uniformly successful requires the closest study and a wide range of remedies.

Aconite—When the different stages are strongly marked. In young persons full of blood. Heat mostly in head and face. During sweat, ear-ache and profuse passing of urine.

Apis—Chill comes in afternoon at 3 or 4 o'clock, worse in a warm room. Chilliness from slightest motion. Thirst during chill; none during sweat. Feet swollen, urine scanty.

Arsenicum—Before the chill: dizziness; yawning; stretching; sleeplessness the night before. Chill may be intermixed with heat; or patient may be cold inside, and burning hot outside. Vomiting during chill if drinks are taken; blue nails. During heat: great thirst, drinks little at a time. Thirst often greatest during sweat. Patient is not well during the intermission. He is weak, pale, and may have a diarrhea.

Bryonia—Great thirst during chill, still more during heat. Cough with stitching pains through chest and abdomen ; pains in the limbs and back, worse from motion.

Quinine—Paroxysms at the same hour, with clear intermissions between. Thirst mostly during sweat. During the paroxysm pain in the spine between the shoulder-blades on pressure. Urine leaves a sediment like brick-dust. Head feels big, with ringing in ears and dizziness.

Eupatorium—Thirst before the chill, which comes on in morning. Nausea, vomiting and great pains in bones.

Hepar—Itching, stinging nettle rash before the chill. Fever blisters around the mouth.

Ipecac—Before chill : yawning, stretching and collection of saliva in mouth ; chill with nausea, vomiting, diarrhea, and difficulty of breathing. Little thirst. During intermission feels bad at stomach.

Opium—Sleeps all through. During sweat feels burning heat.

Rhus tox—Chills come every day, but seem altered every time. Before the chill : yawning, stretching, lame feeling in the jaw. Chill with aching in small of back, wants to lie on something hard, or to press on the back. Pain in the limbs with tingling and crawling in the fingers. Heat with nettle rash. Restlessness.

Arsenite of chinin has often succeeded when there were no special indications for other remedies, or when others failed.

Many cases of ague are reported to have been cut short by giving *Aconite* and *Belladonna* (in alternation) in the commencement of the attack. A trial of this plan can do no great harm.

1. Leucorrhea, *(Whites.)*

This is a catarrhal inflammation of the mucus membrane of the internal genital organs and is characterized by a discharge that may vary from a perfectly clear to a yellowish or greenish color. It is entirely analogous to catarrh of the nasal passages.

This disease is far more prevalent among females of wealth and fashionable methods of living, than among those in the humbler walks of life. And this fact is readily explained by the injurious mode of dress which generally prevails among the class first mentioned. No plan could be devised that would more thoroughly cause stagnation and congestion of blood in the uterine organs, with the result—LEUCORRHEA, than the fashionable style of dress.

The cure of this disease is generally quite easily accomplished, provided the sufferer herself prefers health to fashion.

TREATMENT—The clothing must be worn loose, and should be supported from the shoulders entirely. The skin should be kept in a healthy condition by frequent bathing and daily friction with a flesh-brush. Daily exercise in the open air should be taken, by the best and simplest method—walking. All local applications should be abstained from, excepting warm water for the simple purpose of cleanliness. All astringent injections are harmful although they may temporarily stop the discharge.

Ambra—Bluish-white discharge, more profuse at night. A stitching pain precedes each discharge. Lying down aggravates her symptoms.

Ammonium mur—Leucorrhea like the white of an egg, preceded by a griping pain about the navel.

Calcarea carb—Milky discharge when passing urine, or flowing profusely by spells. Menses too early and too profuse. Feet cold and sweaty.

Hydrastis—Stringy, tough discharge. Great sinking at pit of stomach.

Podophyllum—Thick, clear mucus. Constipation. Bearing down. Falling of womb; bowel protudes at stool.

Pulsatilla—Thin, burning discharge; or it may be thick, white and bland. Menses are late and scanty, or suppressed. Bowels inclined to be loose; chilliness. Timid, yielding, sad women.

Sepia—Women with dark hair and eyes. Yellow color about the mouth and across the nose. Itching in the genital organs. Pains shoot from womb up to navel or pit of stomach. Bearing-down pains; must cross the limbs.

China—This remedy should be used in any case when there is prostration or weakness as a result of the discharge. It is often used in alternation with other remedies. Should be given when there is leuchorrhea instead of menses; with itching. Stomach and abdomen are bloated, yet belching does not relieve.

Graphites—Leucorrhea profuse, perfectly white, especially on rising from bed in morning. Occurs in gushes during the day.

Kali bichrom—Yellow, stringy discharge, with pain and weakness across small of back, and dull, heavy pains in the abdomen.

Secale—Brownish, offensive discharge. Tingling, prickling, or numbness in the limbs.

See following sections.

––––

2. Inflammation of the Womb, (*Metritis.*)

This may occur at any time as the result of

taking cold or violence, but it is more apt to follow
child birth. It sets in with a chill followed by
fever. There is great tenderness of the abdomen,
burning thirst, vomiting, and suppression of the
discharge. It is often accompanied with *milk-leg*,
which is the result of an inflammation of the vein
of the thigh, causing a white, shining, painful
swelling of the whole leg.

TREATMENT—*Aconite*—Fever. Skin hot and dry.
Great fear of dying, predicts the time.

Arsenicum—Burning pains. Sudden sinking of
strength. Burning in the veins. Thirst, drinks
often, but little at a time.

Belladonna—Pains come and go suddenly. Cannot bear the least jar. Delirium. Face red and
hot. Throbbing in sides of neck. Drowsy.
Starts when partly asleep.

Bryonia—Lies perfectly quiet as the least movement causes pain. Stitching pains. Mouth dry
without thirst, or drinks much at a time.

Cantharis—Constant painful desire to pass urine.
Patient lies unconscious with her arms stretched
along side the body; she suddenly starts up,
screams and throws her arms about and may have
a convulsion.   •

Rhus tox—Constant motion; can't lie still.
Tongue dry with red tip. Red rash on the breast.

———

3. Displacement of the Womb.

Displaced forward (*Anteversion.*) *Displaced backward*
(*Retroversion,*) and a *falling* (*Prolapsus.*) May
be caused by a sudden fall, slip or jump,
but in the majority of cases is due to *tight
lacing*, supporting the clothing at the hips,

improper bandaging after confinement, constipation, excessive or long continued dancing, and, lastly, an inherent weakness in the system itself. The symptoms are: a sense of weight and bearing down in the lower part of the abdomen, worse on walking, standing or exertion; pressure upon the rectum and bladder, causing a more or less constant desire for stool or to pass urine, often without success; leucorrhea.

TREATMENT—The sufferer should receive the most intelligent professional aid. A cure can be effected in most cases, but this will require *perseverance* on the part of the patient, for it generally requires a radical change in her habits of life. A few of the remedies used are:

Lachesis—Cannot bear to have the clothing touch the abdomen. Palpitation with numbness of left arm. Constant feeling of something in throat which she cannot swallow down. Feeling of ball rolling in abdomen.

Lilium-tig—Bearing down; pains in left ovary and left breast. Leucorrhea yellow, making the parts sore. Anteversion.

Helonias—Anteverson. Flooding on lifting a weight. Aching and burning in the back. Prolapsus, with profuse flow of clear, light colored urine.

Podophyllum—Falling of womb after overlifting or straining. With falling of the bowel. Pain in lower part of spine. Leucorrhea of thick, clear mucus. Backache after washing.

Sepia—Bearing down pain comes from back to abdomen. Crosses the limbs to prevent protrusion. Sensation after stool as if a lump remained in the rectum.

Pulsatilla—Prolapsus with pressure in the abdomen and small of the back as from a stone;

limbs tend to go to sleep. Women of mild, timid dispositions, and light complexions.

Study section on LEUCORRHEA.

See sections 1, 2, 4.

---

4. Painful Menstruation, (*Dysmenorrhea.*)
Absent or Scanty Menstruation, (*Amenorrhea.*)

TREATMENT—*Viburn op*--This remedy I have found to be a specific almost when the pain comes in paroxysms.

Aconite—Flow ceases from sudden fright or from exposure to cold winds. During puberty frequent bleeding of nose. Bends double on account of pain, but no relief.

Apis—Stinging pains in right ovary. Numbness in right side of abdomen, extending to thigh. Feet swollen.

Belladonna--Hard swelling of ovary; throbbing pain; bearing down as if everything would come out. Delirium, with red face and glistening eyes. Dilated pupils. Rage. Frenzy. Vomiting of blood instead of monthly flow.

Bryonia--Stitching pains worse from least motion. Nose bleed instead of monthly flow.

Cantharis—Constant straining and urging to urinate, with painful discharge of a few drops, which may be bloody. Burning pains in the ovaries.

Cimicifuga—Sharp pains across the abdomen, has to double up. Scanty flow between the menses. Rheumatic pains in the limbs. Spasms at time of menses. Eye-balls ache; very melancholy.

Conium--Menses late and scanty. Shooting pains in breast. Pain from above downward, with drawing pains in legs. Breasts become sore and painful every monthly period.

Colocynthis—Cramp-like pain in left ovary, causing her to bend double; pain in left foot and leg.

Lachesis—Pain and swelling in left ovary; nose bleed before menses. The flow relieves all pains. Can't bear to have the clothing touch the abdomen.

Lilium tig—Burning, stinging, cutting pains in left ovary; pains extend across the abdomen, and down the leg. Bearing down, with pain in left breast. Bright yellow leucorrhea. Pain at the heart as if it was grasped and then released. Very low spirited.

Podophyllum—Bearing down. Sensation as if everything would come out during stool. Numb, aching pain in ovaries; heat down the thigh.

Pulsatilla—First menses are delayed. Too late and scanty. Menses stop from taking cold; getting feet wet.

Sepia—Menses absent or scanty. Great tenderness of female parts. Pains shooting upward to navel and pit of stomach. Yellowness around mouth and across nose.

---

5. Profuse Menstruation, (*Menorrhagia.*)
Flooding, Hemorrhage from the Womb, (*Metorrhagia.*)

TREATMENT—In any serious case, especially when occurring during pregnancy, during or after confinement, no time should be lost in sending for competent medical aid. The foot of the bed should be raised. Injections in the vagina of water hot as can be borne without scalding are said to have stopped flooding when all else failed.

Belladonna—Bearing down. Pain from lower part of backbone through to front. Enlarged pupils. Jerkings of the limbs. Blood bright red, and hot.

Calcarea carb—Menses too soon, too profuse, and last too long. Leucorrhea milky Cold, damp feet.

Cantharis—Menses too early and profuse. Pain ful urging to urinate with scanty flow.

Chamomilla—Drawing pain from small of back forward, followed by discharge of large clots. Tearing pains in the legs. One cheek red and hot.

China—Flow of dark blood; dark clots; spasms in chest and abdomen. Flooding, blood dark; fainting; coldness; ringing or roaring in ears; twitches, jerks; wants to be fanned; convulsions. *China* should also be given after the hemorrhage has ceased, for the debility remaining.

Helonias—Flooding. Pain in back through to uterus. Blood dark. Burning in small of back. Menses too frequent and profuse; prolapsus. Great melancholy.

Secale—Menses too profuse and last too long. Flooding of dark, offensive blood. Cold extremities; cold sweat. Surface cold, yet does not want to be covered. Tingling in back extending to fingers and toes.

Ipecac—Blood bright red; clotted; nausea; heavy sighing breathing; gasps for air; cutting pains from left to right; stitches from navel to womb.

Hamamelis—Steady flow of dark blood.

Trillium pend—Flow of bright red blood.

Sabina—Partly fluid, partly clotted. Pains through lower part of abdomen, from back to front.

Ustilago—Menses profuse, contain small clots. Between the periods. Constant pain under the left breast.

6. Itching of the Genitals, *(Pruritis Vulvæ.)*

TREATMENT—The best external application I know of is *Borax water.*

Calad seguin—Relieves most cases.

Cantharis—With urinary troubles.

Sepia—Leucorrhea. Ringworms on body.

Sulphur—Pimples all around genital organs. Nose itches after menses. Nipples itch.

———

7. Inflammation of the Breast, *(Mastitis.)*

INFLAMMATION OF THE BREAST usually occurs to the nursing mother, or may happen during the period of weaning. It is caused generally by over-distention of the breast from milk ; from pressure of badly fitting garments ; external violence, &c.

SORE NIPPLES are generally the result of THRUSH in the infant, or may be due to a natural tenderness of the skin.

TREATMENT—In any case the result of a blow or bruise, *Arnica tincture* (1 part to 10 of water) may be used externally. When the nipples are cracked and sore, *Calendula tincture* should be used in the same manner. Whenever an abscess forms it should be opened at once, or great harm may be done by the burrowing of the matter or its absorbtion. The infant should not be given the milk of the diseased breast. The milk should be removed with a breast pump. If external applications are made, they must be carefully removed before allowing the child to nurse.

Aconite—When from taking cold. Chilliness or heat. Pulse quick.

Apis—Burning, stinging, swelling of the breasts.

Arsenicum—Burning pains, relieved by motion.

Calcarea carb—Breasts distended, but little milk. Ulcer on the nipple.

Calcarea phos—Child refuses the milk ; milk tastes saltish.   Pain and burning in the breast.   Nipples ache.

Causticum—Nipples sore, cracked, and surrounded with pimples.

Chamomilla—Breasts hard and painful to touch, with drawing pains.   Nipples inflamed and very sore.   The infant's breast is tender to the touch.

Conium—Hard lumps in the breasts with piercing pains, worse at night.   Stitches like needles.

Belladonna—Throbbing pain in breasts.   Bright red in streaks.   Worse in afternoon.

Bryonia—Swelling with little or no redness. Stitching pains worse from motion.   Dizzy and faint when rising.

Phytolacca—Nipples sore and cracked, with great pain when child nurses.   Hard lumps in breast. CAKED BREAST, so called.   Breast swollen and ten-der.   Most cases are benefitted by it.   Abscess forms.

Pulsatilla—Breasts swollen, pains extend to shoulder, neck, chest, and down the arm ; change from place to place.   After weaning, breasts swell, full, tight and stretched ; milk continues to be se-creted.   Hard lumps in the breasts of young girls ; or escape of thin, milky fluid.

Rhus tox—From taking cold.   Inflammation in streaks.   Rheumatic pains worse when at rest.

Sanguinaria—Stitches in breasts ; soreness of nipples, and painful spot under right nipple.

Sepia—Bleeding and soreness of nipples ; they crack very much, across the crown.

Silicea—Breasts swollen, dark red, burning pains prevent rest at night.   Hard lumps.   Great itching of breasts. Burning in nipple. Nipple drawn in. Nipple ulcerates.

Sulphur—Nipples chapped; after warming they smart, burn and bleed. Breast ulcerates, with chilliness in forenoon, and heat in afternoon.

Hysterics, *(Hysteria.)*

A disease depending upon an unnatural sensitiveness of the nervous system and mental or moral nature; together with an inability to control the nervous and emotional manifestations. It is usually, if not always accompanied by some uterine derangement. The attacks are paroxysmal, and with some are apt to occur near or during the menstrual period. There is no absolute uniformity neither in the time nor character of the attacks. They may begin with and consist of spells of immoderate laughing and weeping; or there may be delirium, strange antics, or violent convulsions. A profuse flow of colorless urine is apt to accompany or end the paroxysms.

TREATMENT—Place the patient where there is plenty of fresh air, and loosen the clothing about the neck and chest. Prevent her doing injury to herself in the struggles. All manifestations of compassion or sympathy generally make matters worse. A firm, decided manner is best in dealing with these cases. Any uterine disorder should be corrected, and the general health improved by proper exercise, diet &c. The patient should cultivate the will power, and should try to control her emotions and actions, and she will find her ability to do this to increase with her efforts.

Asafœtida—Sensation as if a ball or lump came

up in the throat, with constant inclination to swallow it down.

Cimicifuga—Hysteric spasms at time of menses. Nervous headache, particularly in forehead and eye-balls.

Caulophyllum—Hysterical spasms at puberty; from menstrual irregularities; with painful menstruation. Moth spots on forehead.

Cocculus—She is absorbed within herself, observes nothing about her. Sadness. Choking sensation in upper part of throat, with difficult breathing and desire to cough. Menses retarded, or too profuse and often. Alternate laughter and crying.

Conium—Sensation of ball or lump in throat. Dizzy when lying down. Breasts become sore and painful at every menstrual period.

Ignatia—Attacks from fright, or grief. Shrieks for help. Choking sensation. Gone feeling at pit of stomach. Frequent sighing. Laughing and crying.

Lachesis—Choking sensation, can bear nothing about the throat, nor pressure about the chest or abdomen. Crawls on the floor, spits often, hides, laughs, or is angry.

Moschus, (Musk)—Suitable for most cases. The odor should be inhaled. Especially where there is frequent fainting, and sensation of constriction of the chest.

Sepia—Sensation as if something twisted about in her stomach and rises toward the throat; her tongue becomes stiff and she is speechless and rigid like a statue. Consciousness remains, yet cannot speak nor stir.

Morning Sickness.

This troublesome symptom of pregnancy may

generally be controlled by a selection from the following remedies: *Anacardium;* when the nausea is worse before and after eating, but better while eating. *Kreosotum;* vomiting before breakfast of sweetish water; breakfast and dinner are retained; vomiting after supper. *Lobelia;* nausea and vomiting, with profuse flow of water from mouth. *Nux Vom;* when there are constipation, bilious symptoms, &c. *Cocculus;* worse when riding in a carriage; can't bear the smell of food. *Petroleum;* worse when riding in a carriage. *Sabadilla;* no relish for food until she begins to eat, when she makes a good meal. *Tabacum;* vomit-with deathly nausea, pallor and coldness; worse from least motion

Abortion. Miscarriage.

Generally due to depressing mental or bodily shocks, to violence, to constitutional weakness, or to a habit of miscarrying. This accident is more apt to occur about the third or fourth month of pregnancy. It sometimes comes on with a sudden, profuse flow of blood from the womb, with faintness and weakness, followed with labor-like pains in the back and abdomen; or it may come on more gradually, with chilliness, languor, loss of appetite, slight pains and uneasy sensations through the abdomen. All discharges should be preserved until the real character of the accident can be ascertained.

TREATMENT—A competent Physician should be summoned at once. The patient should retire to bed (a hard mattress is preferable) at the first symptom of this difficulty. Perfect quietude should be maintained. The room should be moderately cool, and the patient lightly covered, but the feet kept warm. Nourishing broths may be given, and small pieces of ice may be swallowed if there is thirst. The following remedies may be indicated: *Kali carb ;* when there are pains shooting from the back into the buttocks and thighs. *Apis ;* when there are stinging pains in the abdomen. *Aconite ;* when caused by fright, with vexation ; rapid pulse, rapid breathing, fear, and restlessness. *Opium ;* when from fright in the latter months of pregnancy. *Sabina ;* when there is a discharge of bright red blood, partly fluid, partly clotted, worse from any motion, pain through from back to front. *Secale ;* when there are prolonged forcing pains, or discharges of dark, liquid blood, with cramps, numbness and tingling.

See section on Profuse Menstruation and Flooding.

Violent Movements of the Child.

This is often a source of great discomfort during pregnancy. Usually one of the following remedies will lessen this annoyance : *Crocus sat ;* when the movements are accompanied with nausea and faintness. *Opium ;* violent movements, especially after fright. *Sepia ;* when there is soreness of the abdomen ; very sensitive to the movements of child. *Thuja ;* when the movements are so violent as to awaken one out of sleep, and they cause cutting pains in the bladder, with urging to urinate ; or pain in left hip running into the groin.

Nervous Attacks or Shivers.

Occurring during pregnancy or in the first stage

of labor may be controlled by *Gelseminum*, *Cimicifuga or Ignatia* where there is a tendency to sigh.
See section on Hysterics.

After Pains.

When excessive will generally require *Arnica*, especially if they return when the child nurses. *Cimicifuga ;* when the pains are worse in the groins; nausea and vomiting; patient over-sensitive. *Caulophyllum ;* especially after a long, exhausting labor; pains across lower part of abdomen, extending into groins. *Ignatia ;* for after-pains with much sighing. *Sabina ;* when there is much urging to urinate, or the urine may come by drops, with burning; tenderness of abdomen. *Podophyllum ;* when there are heat and flatulence; also strong bearing down. *Secale ;* long-continued after-pains. *Bryonia ;* when the pains are caused by least motion, or deep breathing.

Milk Fever.

This usually occurs at the time of the secretion of milk in the breasts. When excessive it may be controlled by *Aconite ;* when there is a hot, dry skin, thirst, quick pulse, breasts hot, restlessness. *Cimicifuga ;* when there is burning in the breast; or pains below the breasts. *Belladonna ;* when there are heat and redness of the face; throbbing in head; delirium; skin moist and hot.

Suppressed or Difficult Secretion of Milk.

May be the result of taking cold, some mental shock, lack of proper nourishment, or weakness.

Plentiful supplies of milk, rich broths, oat-meal &c., should be taken if there is a lack of nourishment. If the digestion is faulty it should be corrected. Remedies useful are *Aconite ;* when caused by a chill or fright, and the patient is feverish. *Bryonia ;* when the milk is scanty, and the breasts are pale, hard, painful and heavy. *Calcarea carb ;* when the breasts may be full and distended, yet little milk ; patient sweats much about head and neck. *Calcarea phos ;* when the milk is thin and watery; child refuses it, it tastes salt. *Causticum ;* when the milk almost disappears in consequence of over-fatigue, night-watching and anxiety. *China ;* when there is debility resulting from flooding. *Phosphoric acid ;* when there is great indifference concerning herself, or things and persons about her. *Pulsatilla ;* when the milk is suddenly suppressed; the lochia becomes milky, whitish. *Secale ;* when with the lack of milk there is much stinging through the breasts. *Urtica ;* when there is stinging, burning, itching of the skin, like nettle rash ; no milk secreted or it is insufficient.

See section on Inflammation of the Breasts.

———

Excessive Secretion of Milk.

This sometimes becomes a source of great annoyance and exhaustion to the mother. The following remedies will be found useful : *Calcarea carb;* when the breasts are tender and swollen, and the milk runs away of its own accord. Patient sweats much about the head and neck. *China* or *Phytolacca;* when there is an excessive flow with great exhaustion. *Pulsatilla*; when, after weaning milk continues to be secreted; breasts swell, feel stretched, are intensely sore.

See also section on Inflamed Breasts.

———

Derangements of the Lochial Discharge.

This discharge occurs after confinement, and continues generally from 2 to 3 weeks. From a chill, mental shock, getting about too soon, or inflammation of the womb it is sometimes suppressed or changed in character. One or more of the following remedies will be required: *Aconite*; when caused by fright or taking cold; when there is fever, thirst, anxiety, fear, restlessness. *Belladonna;* when the lochia is offensive, and feels hot; or it is suppressed with delirium, red-hot face, staring eyes. *Bryonia;* when the lochia is profuse with burning pain in region of womb. *Cimicifuga;* when suppressed by cold or emotion; or lochia watery, mixed with small clots. *Calcarea carb;* when lochia lasts too long, or has a milky appearance. *Opium;* when suppressed from fright, with drowsiness, stupor. *Pulsatilla;* when the lochia is scanty, milky; feverish but no thirst. *Rhus tox;* when it lasts too long; or returns often; or is offensive. *Secale;* when it becomes dark and very offensive.

———

Change of Life.

This event occurs to most women between the ages of 40 and 50. It may be completed in a few months, and with very little discomfort; or it may occupy many months, and be replete with suffering and peril. The first indication is generally some irregularity in the menses. The flow will be absent or much diminished for one or more periods; it may then occur again for a period or two, to again cease for a longer time perhaps; gradually ceasing in this irregular manner. Most peculiar to this condition are the sudden flushings of the face, the sudden sensations of heat followed by chills or shivers, and the perspirations. The remedies most useful in this condition are, *Amyl nit.: Cocculus ; Lachesis ; Sanguinaria.*

See section on Hysterics.

---

1. Burns, Scalds, &c.

TREATMENT—For simple burns by dry heat, holding the burnt part as near as possible to a hot stove or coal for a little time and repeat, is as good as any thing I know of. It will soon relieve the pain entirely. For scalds or burns where the skin is destroyed the mixture of lime-water and sweet oil is one of the best applications. Its good effects may be increased by adding 2 drops of the *tincture of Cantharis* to 1-2 pint of mixture.

Aconite may be given if there be fever, thirst, &c.

Arsenicum in extensive burns; prostration; diarrhea; vomiting.

Belladonna—Congestion of the head, delirium, &c.

Cantharis—Useful in most cases, especially when there are blisters, or the urine is affected.

Blows, Bruises, and Their Results.

TREATMENT—*Arnica* Internally (1 drop of tincture to a glass of water.) Externally (one part to ten of water.) When the flesh is torn or punctured.

Calendula—Externally (one part to ten of water.) Internally, *Arnica.*

Rhus tox—Strains, Sprains, &c.

---

2. Drowning. Hanging.

In case of drowning first lower the chest and head for a few seconds, that any water in the lungs or air tubes may run out; then lay the person on his back where the air can freely circulate. Loosen the clothing about the neck and chest, and elevate the neck a little. Place the persons arms so his elbows will be nearly or quite as high as his shoulders. Kneeling over the body, looking toward its face, seize the lower part of the chest with the hands expanded, thumbs pointing toward the persons nipples, fingers extending upward and around toward the back. In this position press your hands toward each other smartly and firmly for about two seconds; then immediately withdraw the hands. Repeat this pressure and release every 4 or 5 seconds. An assistant can very materially aid by lowering the persons arms to the sides when the pressure is made and raising them when the pressure is removed. This effort may be continued for two hours. At the same time brisk rubbing and heat may be applied to the lower part of the body. If the least sign of life is manifested the efforts should continue until breathing is fully established.

The same process in *hanging* or strangulation from any cause.

Tart emetic—May be given as soon as the person shows returning life; a drop or two being placed on the tongue.

Freezing.

In a case of apparent death from extreme cold, the body should be placed in a cold room and covered with snow, or bathed in ice- water until the limbs are soft and pliable, then it should be placed in bed and briskly rubbed with dry flannel, trying at the same time to induce respiration by the method given in case of drowning. A little coffee without milk may be given by injection as soon as there are signs of returning life, and spoonful doses as soon as he can swallow. In all cases of freezing snow or cold water should be applied constantly until the *frost is drawn out*; afterward the frozen part will be apt to exhibit the symptoms of a burn and should be treated as such.

---

3. Profuse Bleeding.

from a wound should be controlled by tightly cording the limb above the wound. This may be done with a piece of rope or handkerchief and advantage gained by twisting with a stick. If the artery from which the bleeding takes place can be felt, a knot in the cord or handkerchief should be placed over it. Bleeding from the arteries in the extremities can be controlled to quite an extent by flexing the limbs—that is, bending them sharply upon themselves. If the bleeding should be from a wound in the arm below the elbow, the arm should be bent up across the chest, and the forearm tight up against the arm. So in the lower extremities, the thigh may be bent as close to the body as possible, and the leg upon the thigh. This method creates sharp turns in the artery, thus impeding the flow of blood considerably.

---

4. Poisoned Wounds.

These should be immediately and thoroughly *sucked* until a cord or handkerchief can be tightly tied a few inches above the wound. Strong carbolic acid should be thoroughly applied if it is at hand ; or the wound may be cauterized by nitrate of silver, or one of the strong mineral acids (nitric, sulphuric, muriatic) ; or a red hot coal, or the lighted end of a cigar, or the end of a heated iron should be held close to the wound ; and this application of dry heat must be constant for some time

---

Poisoning.

TREATMENT—When the poison has been swallowed, the first object is to get it out of the stomach. Vomiting should be excited by tickling the throat with a feather, or something similar; or a little salt and snuff, or salt and mustard may be placed upon the tongue. The most efficient emetic is *sulphate of zinc*, of which 24 grains, dissolved in a little water, may be given to an adult, and repeated if necessary. Tobacco smoke blown into the rectum by means of a common pipe, is often efficient to produce vomiting. *Ipecac* and *ground mustard*, in 30 grain doses are efficient emetics, and the latter is usually at hand.

These means should be aided by plentiful draughts
of warm water. When available, the stomach-
pump should be used.

POISONING BY ACIDS—(Muriatic; Nitric; Sulphuric.) Give warm soap-suds and milk; or
magnesia in water; or powdered chalk in warm
water; or soda or saleratus-water; or wood-ashes
in water. Afterward give mucilaginous drinks,
such as slippery-elm water, or flax-seed tea, or
white of egg stirred up in water.

For after effects see Gastritis.

POISONING BY ALKALIES—(Ammonia or Hartshorn; Potash, Washing Compound; Sal-ammoni-
ac; Lye.) Do not vomit. Give dilute vinegar; or
lemonade; or sour milk or butter-milk; or sweet
oil, or linseed oil, or castor oil, or melted tallow
or lard; afterward mucilaginous drinks.

POISONING BY ANTIMONY—Give strong decoction
of nut-galls, or oak-bark; or strong black tea; or
strong coffee; mucilaginous drinks.

POISONING BY ARSENIC—(Fowler's solution; Fly-
powder; Paris-Green; Cobalt.) Produce vomiting or use stomach pump. For *arsenic* itself, the
best antidote is the *peroxide of iron,* a teaspoonful
every 10 minutes until relieved. If this is not at
hand use iron-rust in sugar-water. For Fowler's
solution, lime-water may be given, freely. Sugar-
water; or white of egg-water; or linseed tea; or
magnesia in water; or soap-water may be given.
Afterward mucilaginous drinks; and see *Gastritis.*

POISONING BY CANTHARIDES, (*Spanish flies.*) —
Vomit, then give white of eggs, mucilaginous drinks,
sugar, water, milk, &c. Afterward, use camphor
in small doses. For the poisonous effects upon
the urinary organs, from swallowing the poison or
from blister plasters, *see sections under urinary organs.*

POISONING BY COPPER.—(Blue Vitrol, Verdigris,

Paris Green, Food cooked in copper vessels, Pickels made green in brass kettles.) Vomit, then give white of eggs, sugar water, milk, and mucilaginous drinks. Vinegar should *not* be given. Afterward, treat for *Gastritis*.

POISONING BY MERCURY—(Corrosive sublimate, Bed-bug poison, White precipitate, Red precipitate, Vermilion red.) Vomit, give white of egg-water freely; or sugar-water; or starch boiled in water; or flour-paste and milk freely. Afterward treat for *Gastritis*.

POISONING BY GASES–(Carbonic acid gas; Ether; Chloroform.) Expose the person to the fresh air. Bathe the face and chest with vinegar and give strong coffee if the person can swallow. If necessary, resort at once to the method for inducing artificial respiration, explained under *drowning*.

POISONING BY LEAD—(Sugar of lead; White lead; Red lead; Litharge.) Vomit. Give a solution of Epsom Salts; or white of eggs; or soap-suds; or milk.

POISONING BY SILVER–(Nitrate of silver or Lunar caustic.) Give common salt dissolved in water, in large quantities. Afterward mucilaginous drinks.

See Gastritis.

POISONING BY NARCOTICS. (Opium or morphine; Aconite; Oil of bitter almonds; Belladonna or Deadly Night-shade; Cicuta or Hemlock; Camphor; Stramonium or Thorn apple; Digitalis or Foxglove; Hyoscyamus or Henbane; Tobacco; Nux vomica; Strychnia; Rhus tox or Poison Oak or Sumach; Ergot or Smut Rye; Veratrum Album or White Hellebore; Conium or Poison Hemlock; Poison Mushrooms &c.) Vomit by mustard or sulphate of zinc; repeated; or stomach pump. If there is drowsiness or stupor, keep the

patient aroused by constant motion, whipping &c.
Artificial respiration may be resorted to when
necessary. The best direct antidote for opium or
morphine, is injecting *Sulphate of Atropia* under
the skin. The thirtieth of a grain may be used at
once, and repeated every half-hour until relief
occurs. For the others *strong coffee* is the best
antidote and is also valuable for opium. To the
coffee may be added a few drops of hartshorn;
this given every 15 or 20 minutes. If the poison
has been expelled, vinegar and water may be
given in alternation with the coffee. Galvanism
or Electricity is a valuable agent, and may be
used by applying the *positive* pole to the inside of
the mouth and the *negative* pole to the lower part
of the breast-bone. When spasms occur they may
be controlled temporarily by the inhalation of
chloroform.

POISONING BY ACRID VEGETABLE SUBSTANCES—
(Arum Tryphyllum or Indian Turnip; Croton
Tiglium or Croton Oil; Sabina or Oil of Savine;
Phytolacca or Poke; Podophyllum or Mandrake;
Tansy &c.) Treatment as in Poisoning by Nar-
cotics.

POISONING BY IODINE OR IODIDE OF POTASH—
Vomit. Give starch-water; or flour-paste; linseed tea &c.

POISONING BY PHOSPHORUS—Vomit, then give
magnesia stirred in water. Mucilaginous drinks.

POISONING BY PRUSSIC ACID—Inhale the spirits
of hartshorn and weak solution of same internally. Pour cold water upon the spine from a height.
Strong coffee to drink and as an injection.

POISONING BY ZINC—(Acetate of zinc; sulphate
of zinc, white vitrol.) Give soda or saleratus in
water. White of eggs, milk &c.

POISONING BY CREOSOTE; OIL OF TURPENTINE &C.
—Vomit and give white of egg.

POISONING BY POISONOUS FISH, PORK OR SAU-
SAGE—Vomit. Give a brisk purgative. Afterward vinegar and water.

POISONING BY BITES OF INSECTS—Hartshorn may
be rubbed into the part. A weak solution may be
taken internally, in drop doses. Also *Camphor*
may be used in same manner.

See poisoned wounds.

POISONING BY MUSHROOMS—Produce free vomiting by giving mustard, cold water, tickling throat
&c. Afterward give powdered charcoal mixed
with sweet oil. Smelling of hartshorn is also beneficial ; strong coffee.

Remarks Upon Antidotes.

Acids, such as dilute vinegar, lemon-juice &c.,
are useful in poisoning with the alkalies; but they
should not be used in cases of poisoning by arsenic, or vegetables with an acrid juice, such as mandrake, poke-root &c.

Coffee, strong and black, and taken hot as possible is valuable in most cases where there is drowsiness, intoxication, loss of consciousness, delirium,
&c. Against narcotics in general.

Camphor is useful in poisoning by vegetable or
animal substances, especially where there is vomiting, purging, pale face, cold extremities, and loss
of consciousness. In poisoning by venomous insects, it is also valuable.

Mucilaginous drinks, such as gum arabic water,
slippery-elm water, flax-seed tea, &c., are useful in
poisoning by alkalies, given in alternation with
acids.

Sweet oil is far less useful than generally supposed. It is injurious in poisoning by arsenic, or against the bad effects of spanish flies.

Soap Water is one of the most useful in poisoning by the acids. Common soap may be dissolved in about four times as much water; a teacupfull of this may be given every few minutes.

Sugar water is useful in poisoning by acids or alkalies, after the direct antidotes have been given. It is very useful in poisoning by colors, verdigris, copper, alum &c. Useful also against arsenic, and acrid vegetables.

White of Egg stirred in water is one of the most useful remedies against poisoning by metallic substances, chiefly against corrosive sublimate, mercury, verdigris, tin, lead, sulphuric acid. It should be given in any case, when the poison is unknown, and there is great pain in stomach or abdomen, with violent urging to stool; or diarrhea with burning and pain in the anus.

1. Aconite.

Usually the first remedy to be used in all cases characterized by feverishness unless some other remedy is plainly indicated. Dryness of the skin, thirst, restlessness, fear of death, heat of the body, or a chill from the blood being driven from the surface to some internal organ, are indications for its use. It is very usetul in the early stage of most cases of cold, catarrh, sore throat, influenza, acute rheumatism, or inflammation of any of the organs. It affects most readily persons of full habit, who are full of blood, who have dark hair and eyes. *Aconite* is antidoted by vinegar.

SPECIAL INDICATIONS—Delirium at night. Fear of a crowd. Fear of death, predicts the time. Dizziness when raising the head. Fainting when raising up with paleness of face. Fullness in head. Crampy sensation at root of nose. Headache as if the brain were moved, or raised. Burning, as if brain were moved by boiling water. Headache with increased secretion of urine. Roaring in ear. Cannot bear music, makes her sad. Face red and pale alternately ; one red, one pale cheek. Face feels as if grown larger. Complaints of upper lip. Burning, tingling and numbness of lips. Teeth sensitive to cold air. Everything tastes bitter except water. Tongue feels as if swollen, burning, prickling and numbness. Numbness of mouth. Numbness and burning in throat. Desire for wine or brandy, beer or bitter drinks. Ice water causes cough. White stools with dark red urine. Alternate slimy stools and constipation in jaundice. Profuse urination with headache and sweat. Children take cold and can't make water. Uterine hemmorrhage, much excited, dizzy, can't sit up, fear of death. Hot feeling in chest. Bleeding from lungs. Bruised pain between the shoulders. Crawling sensation in the spine. Numbness in the small of back. Numbness of arms, hands and fingers. Shooting pains in all the limbs. Coldness of hands and feet, paleness and soles sweaty. Body hot, extremities cold. High fever, great thirst, restlessness. Pulse quick and hard. Sleepless after midnight. Sleepless from fear; fears he will die. Profuse hot sweat during sleep. Bad effects from getting cold when sweaty; from cold, dry weather.

Aethusa.

Very useful in the disorders of infants, especially those characterized by spasms or convulsions, or a tendency thereto. There is usually squinting, staring, dilated pupils, rolling of the eyes, or they may be turned downward. The principal indications for this remedy is the *vomiting of milk*. This is apt to be thrown up in curds.

SPECIAL INDICATIONS—Distressing pain in back of head, nape of neck, and down the spine, better from bending stiffly backward. Head feels as if in a vice. Objects look larger than natural. Eyes brilliant, protruding. Face puffy, spotted red. Chin and corners of mouth feel cold. Tongue feels too long. Taste in the mouth like cheese, like onions, sweetish in morning. While eating sudden heaviness in forehead. Sudden vomiting of white, frothy substance; of curdled milk; of green phlegm. Food comes up an hour or so after eating. Coldness of abdomen and lower limbs. Bluish-black swelling of abdomen. Green, slimy stools; of undigested food. Pain in kidneys, worse from sneezing and from breathing deep. Cutting pains in bladder, with frequent calls to urinate. Red urine with white sediment. Right testicle drawn up with pain in kidneys. Small of back feels as if in a vice. Numbness of arms; they feel as if shorter. Spasms with clenched thumbs. Eyes turned down, teeth set, foam at mouth. Black and blue spots all over.

Arsenicum.

This remedy will be found useful generally, in cases of disease characterized by great prostration. There will be great restlessness also, perhaps, yet the patient so weak that he can only move his limbs about, his trunk being quiet. The circulation of the blood is apt to be languid. The extremities are cold, and the feet are inclined to bloat. The pains in the stomach and bowels are of a burning character, while the pains in the limbs are of a drawing, tearing kind. The thirst is peculiar; there is generally a want of but little drink at a time, but it is required often. Diseases calling for *Arsenicum* are apt to be worse at midnight or just after; also worse from cold generally.

SPECIAL INDICATIONS–Great anxiety and restlessness. Fear of death or that he will never recover. Cannot find rest anywhere, wants to go from one bed to another. Heaviness in head, and humming in ears, goes off in open air, returns in room. Pain in left side of head and face. Throbbing headache over root of nose. Head very sensitive to open air; must wrap the head up warmly. Sunken eyes. Burning in eyes. Trembling of upper eyelids. Roaring in ears with paroxysms of pain. Fluent catarrh, sneezing, horseness, swollen nose. Watery discharge causing burning and smarting at nostrils. Distressing stoppage at bridge of nose. Face pale, sunken, deathly ; waxy, puffy. Tearing in left-half of face. Burning, stinging pains as from red hot needles. Pimples, itching and burning worse at night and in cold air; better from warmth. Food tastes too salty. Root of tongue swollen. Aversion to butter. Drinks often, but little at a time. Thirst, but no desire to drink. Vomiting of food and drink immediately after taking it. Great burning in stomach with vomiting of greenish-yellow, brown, black or bloody fluid. Spleen enlarged. Burning in abdomen. Diarrhea from chilling the stomach. Stools like dirty water; foul smelling. Purging, with cold extremities. Anus red and sore. Burning in anus. Piles with stitching pains when walking or sitting, not when at stool. Bloody urine. Thin, whitish, offensive discharge in place of menses. Leucorrhea, profuse, yellow, thick, makes the parts sore. Cough preceded by jerking in hips which seems to cause the cough. Asthma, worse in stormy weather and heavy air. Stitches in breast bone from below upward when coughing. Knees crack when walking. Itching of feet and thighs. Uneasiness of lower limbs, cannot lie still. Great thirst during sweat. Profuse sweat about the knees. Dropsy of abdomen ; of feet, of face. Eruptions with burning-itching, parts pain after scratching. Worse in general about midnight. Better from heat generally.

Belladonna.

This remedy is especially suitable for persons with light hair, blue eyes, fine complexion and delicate skin; or persons full of blood, with fevers and inflammations. Skin becomes exceedingly hot, may be dry or moist. Suits well persons who are plethoric, who, when well, are jovial and entertaining, but are violent when sick. It will be used principally in disorders of the head and throat. In the head troubles there will usually be a sensation of fulness and beating, throbbing in the sides of the neck, head hot, face red and pupils dilated. The least jar aggravates the pains. In the throat troubles there are soreness, redness, dryness and heat; a constant sensation of something in the throat that must be swallowed down; right side apt to be worse.

SPECIAL INDICATIONS :—

Delirium; sees ghosts, monsters, insects &c. Desires to hide or escape. Picking at bed clothes. Feeling in head like swashing water. Cold feeling in middle of forehead. Fulness and pressure in head; pain in head and eyeballs, eyes feel as if starting from their sockets. Headache worse when lying, better when sitting quiet. Beating in head. Pains come suddenly and cease suddenly. Headache from heat of sun. Objects seem double or inverted. Pupils dilated. Eyeballs red and glaring. Nose bleed with congestion of head. Nose bleed in children at night. Sensation of burning heat in face without redness of cheeks. Shooting from side of neck up to ear and temple. Upper lip swollen; cracked in middle. Sensation as if lower jaw was drawn backward. Teeth feel on edge. Gums swollen on right side. Taste salty or sour. Bread tastes sour. Tongue with white centre or two white strips. Mouth feels hot. Throat raw, sore; looks red and shining. Constant desire to swallow. Throat worse swallowing liquids. Pain in stomach compels patient

to bend backward. Clawing pain in abdomen.
Stools contain lumps like chalk. Involuntary
stools. Sensation of a worm in the bladder.
Wets the bed, starts and jumps in sleep. Profuse
discharge of hot, bright red blood. Asthma in
hot, damp weather; worse after sleep. Quick,
short breathing alternates with slow, gentle
breathing. Violent cough about noon. Cough
ends with sneezing. Child cries before the cough.
Cough at 11 P. M. Backache as if broken.
Throws body first forward and then backward
when lying. Convulsions commence in arm.
Starts up in fright during sleep. Boils in spring.
Bright red swellings anywhere. Skin imparts a
burning sensation to hand. Skin scarlet and
smooth. Red streaks extend from centre of swelling. Young persons, full of blood; children with
florid skin, bright hair, blue eyes.

Bryonia.

A marked characteristic in disorders calling for
this remedy is the aggravation of the suffering
from motion. It is useful in rheumatism and
headache when this peculiarity is present. The
thirst of *Bryonia* is for large draughts of water,
but not so very frequent. The mouth is dry with
a bitter taste; the tongue thickly coated, white or
yellowish. In chest troubles the pains are sharp
and stitching, worse from breathing, coughing or
motion. Sour stomach, with a sensation of a
weight or load after eating.

SPECIAL INDICATIONS :—

Dizzy as though the brain was turning around;
when rising or raising the head. Pressure in the

brain forwards. Headache commences when opening the eyes in the morning. Digging pressure in forehead, worse when stooping or walking fast. Morning headache. Hair feels greasy from sweat. Pain in eye, worse on moving it with feeling as if eye ball became smaller and was retracted within the orbit. Twitching of left upper eyelid with heaviness. Chirping in head as from locusts. Nose bleed in morning after rising; sometimes during sleep at 3 A. M. Swelling of tip of nose with twitching pain in it. Upper lip and nose swollen, red and hot. Cracks in lower lip. Constant motion of mouth, as if chewing; in brain affections. Cold water relieves tooth ache. Mouth tastes bitter. Tongue coated thick, white or yellow. Much soapy, frothy saliva in mouth. Offensive smell with hawking round cheesy lumps size of pea. Desire for things to eat; when offered they are refused. Desire for coffee, wine. Eats little and often. Drinks much at a time but not so often. Nausea from slightest motion. Vomiting of food but not of drinks. Abdomen distended, yet stomach feels empty. Pain in stomach worse from coughing or motion. Diarrhea; stool passes by a single prolonged effort. Constipation; stools large, hard and dry. Aching piles. Breasts are hard, heavy, pale and painful. Sore mouth of infant; child does not like to take hold of the breast, but after the mouth becomes moistened it nurses well. Cough with crawling and tickling in pit of stomach, worse after eating and drinking and coming into a warm room, with vomiting of food. Stitches through chest on coughing or breathing. Cramp in region of heart, worse from motion. Stiff neck. Stabbing pains in hips. Pains in limbs, worse from motion. Inner side of knee very sore. Instep hot and swollen. Limbs heavy, like lead. Faint when rising from bed; from least motion. Spasms when measles strike in. Chill frequently on right side only. Sour, greasy sweat.

Cantharis.

In most disorders where this remedy is useful there will be found more or less irritation of the urinary organs. Burning and smarting when passing urine, with an almost constant or a frequent desire to urinate, and passage of but few drops at a time, characterize this remedy. It may be used in any case when the above symptoms are prominent.

SPECIAL INDICATIONS :—

Sudden loss of consciousness, face red (during teething). Paroxysms of rage, with crying, barking and beating; renewed by sight of dazzling, bright objects, or when trying to drink. Sensation on top of head as if a lock of hair were being pulled. Hot sensation passes out from ears. Tearing in and behind right ear. Erysipelas with blisters. Blisters on tongue. Small blisters in mouth. Throat feels as if on fire. Drinking increases pain in bladder. Vomiting of blood with burning in stomach. Great bloating of abdomen. Burning in anus after stool. Desire for stool while vomiting. Stools blood and mucus. Pain in kidneys. Pain, burning and smarting before, during and after urinating. Pains in bladder. Urine passes drop by drop. Sweat smells like urine.

Cina.

SPECIAL INDICATIONS :—

Pitiful crying. Child does not want to be touched, nor that you come near it. Desires many things which are refused when offered. Empty, hollow feeling in head, with inclination to vomit. Headache, pain in chest and back, from fixing eyes steadily on some object, as when sewing, &c. Feeling of weight in top of head. Child leans head sideways; turns head from side to side. Child rubs the nose, or bores the nose in the pillow. White and bluish around the mouth. Grinds the teeth. Hiccough during sleep. White stools. Urine milky. Frequent motion as if swallowing something. Gagging cough. Epilepsy, with full consciousness and stiffness. Sudden distressing cries during sleep. Before chill and after sweat, vomiting of food, with canine hunger at same time.

Colocynthis.

SPECIAL INDICATIONS :—

Pressing, burning pain in left eye, temple and bridge of nose. Bad effects from anger, especially vomiting and diarrhea. Headache in front, worse stooping and lying on back. Pain in head when moving upper eye-lid. Left sided neuralgia. Diarrhea after least food or drink. Potatoes cause pain in abdomen. Cramping pains in abdomen, relieved by pressure and bending double. Colic; pains paroxysmal; has to press abdomen against something for relief. Diarrhea from least food or drink with colic pains. Stools like scraping of bowels. Urine thick like glue-water. Crampy pain in left ovary. Pain causes her to double up.

Violent pains in thumbs. Drawing, cramping pains in legs. Cramps in muscles. Sweat smells like urine.

Hepar.

SPECIAL INDICATIONS:—

Boring headache at root of nose. Aching in forehead from midnight till morning. Scalp sore to touch, especially after mercury in large doses. Moist eruptions on head. Boils on head and neck. Lumps on head sore to touch. Eyes protruded, (in croup). Itching of outer ear. Nose-bleed after singing. End of nose red and swollen; with a cold or cough. Bridge of nose sore. Itching around the mouth. Upper lip has sore pimples on border; cracked through middle. Sores at corners of mouth. Sensation of plug or splinter in throat. Desires sour, sharp or strong tasting things. Frequent short spells of nausea. Waterbrash. Colic, with dry, rough cough. Constipation, especially when there are eruptions in bend of elbows, and under knees. Stools not hard but difficult to expel. Diarrhea; child has a sour smell; stools smell sour. Wetting bed at night. Profuse menses in women with chaps and cracks in skin of hands and feet. Cough, choking; with swelling below throat pit. Cough caused by a limb getting cold. After cough, sneezing and crying. Offensive sweat under the arm. Tumor at the point of elbow. Palms of hands itch. Hard, burning lumps. Glands inflame, swell and suppurate. Burning and itching with white vesicles after scratching. Eruption in hollow of knee and bend of elbow. Ulcers, with little pimples around them. Stinging and burning in edges of ulcers. *Hepar* is valuable after a person has taken two much mercury or iodide of potash.

Ipecac.

SPECIAL INDICATIONS :—

Headache down to root of nose, with nausea and vomiting; better in open air. Blue and red rings around the light. Red circle around the cornea. Shooting pains through eye-balls. Ears cold during fever. Eyes sunken with blue margin, face pale. Skin around the mouth red. Lips blue during chill. Child thrusts his fist into his mouth; screams; during dentition. Troubles from rich cake, fat food, pastry, ice-cream, &c. Nausea constant, with almost all complaints. Vomiting with many complaints. After vomiting, sleepy. Stomach feels as if hanging down. Cutting pains in bowels running from left to right. Diarrhea with nausea, vomiting, especially in fat, pale children. Cough constant, no phlegm yields, though chest seems full. Strangling—so much phlegm seems to accumulate in the wind-pipe. Rattling of large bubbles; cough and gagging. One hand cold, the other hot. Spasms from tobacco. On falling asleep, shocks in the limbs. Sweat smells sour; stains yellow. Worse during sweat, better after.

Kali Bichromicum.

Ill-humored; low-spirited. Head light across the fore-head on stooping. Dull heavy throbbing above the eyes, after dinner. Headache preceded by blindness.

Dull heavy throbbing above the eyes after dinner. Headache preceded by blindness. Shooting pains from root of nose over left eye. One-sided headache in small spots. Small brownish or reddish spots on white of eye. Stinging and beating in ears; discharge from ears. Lobe of ear itches. Loss of smell. Discharge from nose, stringy; tough green masses. Scabs or ulcers on septum of nose. Sensation as of a hair up in nose. Face red in blotches. Shootings in left upper jaw-bone toward ear. Lower lip swollen and chapped. Mumps on right side. Sensation of hair on back part of tongue. Deep sore on edge of tongue. Deep sores on palate. Sharp pains in left tonsil. Longs for beer; for sour drinks. Aversion to meat. Stitches through abdomen extend to backbone. Diarrhea, watery, gushing, in morning. Dysentery every year, early part of summer. Urine has a white film and a white sediment. Hawking of copious, thick, blue mucus in morning. Croup worse from 2 to 3 in the morning; tough phlegm strangles him. Sensation of choking when lying down. Cough with phlegm that can be drawn in strings to the feet. Pains from back to breast-bone. Pains in shoulders, worse at night. Pain from left hip down to calf of leg. Soreness in heels. Sweats on back during stool. Pains fly rapidly from one place to another. Pains in small spots. Brown spots on skin. Ulcers deep, with over-hanging edges. Affects most promptly light-haired persons; fat, chubby children.

Lachesis,

This remedy will be found useful for many of the ailments of females occurring at the change of life. The flushings, sudden sensations of heat, tremulousness, nervous palpitations, and choking sensations so apt to occur at that time are generally relieved by *Lachesis*. It is useful in throat troubles, especially when upon the left side, or worse upon that side. There is usually a strangling, choking sensation when lying down, or after having been asleep. It will afford great relief in hay-fever when this strangling sensation is present. Patients requiring *Lachesis* are generally very sensitive about the throat, can't bear anything tight about the throat.

SPECIAL INDICATIONS :—

Delirium, with great talkativeness, constantly changing from subject to subject. Delirium-tremens, worse after sleep; cannot bear anything about the neck. Thinks she is dead; fears the medicine is poison. Talks, sings, whistles, makes odd motions. Great sadness in morning when awaking. Dread of death, fears to go to bed. Burning in top of head at change of life. Feeling in back part of head as if pressed apart. Sneezing in hay-asthma. Yellow complexion with extreme redness of cheeks. Neuralgia, left side. Tongue trembles when put out, or catches behind the teeth. Red tip and brown centre. Sensation in roof of mouth as if it was peeling off. Feeling of lump in throat; seems to go down when swallowing, but returns at once. Diphtheria. Quinsy, &c. Great tickling in throat produces a strangling, especially when lying down; in hay-fever. Desire for oysters; for liquors; for coffee. Symptoms worse after acids. Fulness after eating; belching relieves. Pit of stomach painful to touch. Gnawing and pressure relieved after eating but returns when stomach becomes empty. Can't bear any pressure about the waist. Abdo-

men bloated, can bear no pressure. Stools of de-
composed blood, looks as if made up of pieces of
charred straw. Diarrhea worse at night and after
acids. Itching in anus, worse after sleep. Beating in anus. Piles. Urine almost black. Troubles in left ovary. Flushings at change of life.
Throat very sensitive to least touch, can bear
nothing around it. Gagging cough from tickling
in throat, or in chest, or at pit of stomach ; worse
when falling asleep. Great difficulty of breathing
when awaking. Numbness, tingling and prickling in left arm and hand. Finger tips numb in
morning. Sciatica, left side ; pain as from a hot
iron ; worse after sleep. Aching in shin-bones.
Ulcers on legs, with blue border. Erysipelas ;
surface blue and glossy. Epilepsy comes on during sleep. Left-sided paralysis. Trembling all
over. Awakens at night, cannot sleep again.
Sweat stains yellow or red. Small wounds bleed
much. Ulcers with blue or purple border, and
small pimples around them. Carbuncles. Bed-
sores with black edges.

———

Lycopodium.

SPECIAL INDICATIONS :—
Absent-minded, thinks he is in two places at the
same time. Doubts about one's salvation. Dizzy
when drinking. Headache worse from 4 to 8 P.
M. Eruptions begin on back of head ; discharge
thick. Dandruff in spots. Sparks before the eyes
in the dark. Night-blindness, with black spots
before the eyes. Sees only the left half of an object with distinctness. Sensation as if hot blood
rushed into the ear. Polypus of the ear. Nose

stopped up at root, breathes through the open mouth. Snuffles, child starts out of sleep rubbing its nose. Fan-like motion of the wings of nose. Copper-colored eruption on forehead. Face pale, with circumscribed red cheeks. Deep furrows in cheeks. Lower jaw hangs down during sleep. Eruption around the mouth. Ulcer on the red border of the lower lip. Gum-boils. Ulcers under the tongue. Putrid smell from mouth in the morning. Sore throat from right to left. Desires sweets, oysters; aversion to bread, coffee, meat, warm, boiled food. Great hunger, the more he eats the more he craves. Hungry, but a few swallows fills him up. After eating, fulness, drowsiness. Spitting up food. Sour taste; sour belchings. Belching affords no relief. Nausea when riding in a car or carriage. Tightness about the waist as from a hoop. Rumbling of wind under left short ribs. Colic in right side of abdomen, extends to bladder, with frequent desire to urinate. When turning on right side, a hard body seems to roll from navel to that side. Great quantity of wind in abdomen. Brown spots on abdomen. Backache relieved by passing urine. Pains in bladder worse at night; or when lying down; better from horse-back riding. Soreness in the groins and on the thighs. Leucorrhea, milky; or bloody, red; worse before the full moon. Labor pains run upward. Hard, burning lumps in the breasts, with stitching pains. Hoarseness after croup. Cough worse from 4 to 8 P. M. Brown spots on the chest. Chest sounds full of mucus; raises mouthfuls of mucus, of a light rust color. Large clusters of red pimples on neck. Burning between shoulder-blades. Cramp in calves and toes when walking, Swelling of the soles. Bad smelling foot-sweat with burning in soles. One foot hot the other cold. Sleepy during day, wakeful at night. Very cross when awaking. Chill from 4 to 8 P. M. Nausea and vomiting, then chill, then sweat, without heat. Sour vomiting after chill, also hands and feet bloat. Dard red blotches here and there. Flesh in ridges as if struck with a stick. Blood-boils. Boils which do not come to a head.

Mercurius.

Will be found useful, principally in disorders o teething, sore mouth and dysentery. In teething and sore mouth there will be, generally, a profuse flow of saliva. In dysentery the discharges are bloody and slimy, with a constant urging and feeling as though one could not "get done." The patient requiring *Mercurius* will generally be worse in the evening and during the night, and in damp, cold weather. He is apt to sweat with most complaints, but seems to get no relief from sweating.

SPECIAL INDICATIONS:—

Homesickness. Sweetish risings in throat, then fainting followed by sleep. Dizziness when lying on the back ; when stooping. Burning in left temple, worse at night lying in bed. Head feels as if in a vice, with nausea; worse in open air. Tightness over forehead as if in a hoop. Head feels as if it would burst, as if it was getting larger. Bony enlargements on the skull. Sour, oily sweat on the head, worse at night. Black points before eyes. Sore eyes, profuse, burning discharge, worse at night; pimples on the cheeks. Boils in the ear. Constant cold feeling in the ears. Nose bleed when coughing and during sleep; blood hangs in a dark string. Sneezing; fluent discharge; nose red and swollen; worse in damp weather and at night. Swelling of side of face with heat and toothache. Pimples; bluish red around; do not itch. Toothache; worse at night. Tongue red, with dark spots. Moist tongue with great thirst. Tongue swollen, flabby, takes the prints of the teeth. Sore mouth. Great flow of saliva. Coppery taste. Sore throat. Stinging pains; worse at night; from empty swallowing. Hunger even after eating. Nausea with a sweet taste in the throat. Jaundice. Abdomen hurts when lying on right side. Bloody, slimy stools; much straining and griping; but little passes; feels as if one could not get done. Piles. Leucorrhea, contains lumps; worse at night. Pain in the breast at every monthly period. Milk in breast instead of menses. Inflammation of breasts. Cough as if head and chest would burst, sometimes with vomiting; worse at night. Palpitation; worse at night. Red, hot swelling from elbow to wrist. Back of the hands raw, cracked; cracks on the joints; burning, stinging pains. Cold, clammy sweat on legs at night. Sleepy during the day, sleepless at night. Convulsions. Chill at night with frequent urinations. Sweats at night, with burning of the skin. Often feels worse during sweat. Eruptions. Shingles. Generally complaints worse at night and in damp weather.

Nux Vomica.

Useful in disorders of the stomach and bowels, especially when occurring in thin persons, with dark hair, who are irritable and get angry easily;

also persons who are in-doors much, or study much; who are addicted to the use of coffee and liquors; in those who are inclined to be sleepy in the evening, and to awake at 2 or 3 in the morning, to remain awake for two or three hours, and then to sleep late in the morning. The symptoms come on or grow worse early in the morning, and are aggravated by eating or mental effort. *Nux vomica* and *Sulphur* in alternation, are a favorite prescription with some for piles.

SPECIAL INDICATIONS :—

Irritable; morose; melancholy. Depressed mood in those who dissipate, or are of sedentary habits. Over-sensitive to light, noise, sounds, smell; every little thing offends. Ailments after mental labors. Dizzy in morning. Burning in forehead in morning. Pain in back of head, with dizziness and deranged stomach. Scalp sensitive to touch or wind; feels better wrapped up. Light hurts the eyes; worse in the morning. Blood red spots in the white of the eye. Lower half of eye-ball yellow. When swallowing pushing out pain in ear. Catarrh dry at night, fluent by day. Face yellow with red cheeks. Neuralgia in the face with water from the eye and nose of the affected side. Face feels numb. Bad taste in mouth in morning. Longing for fat food or chalk. After eating; sour taste; waterbrash; pressure in the stomach; must loosen the clothing about the waist. Hiccough Indigestion from sedentary habits; abuse of drugs; abuse of liquors; night-watching; after too high living. Stomach pains worse before breakfast. Jaundice. Colic before breakfast. Alternate constipation and diarrhea. Constipation with rush of blood to the head; with frequent fruitless urging. Piles. Pain in right kidney, extending to genitals and right leg; worse lying on right side. Bloody urine from suppressed flow from piles, or from suppressed menses. Pressure, weight, and bearing down, worse in the morning. Prolapsus uteri from lifting or straining. Menses too early and profuse; faints easily. Whooping cough. Palpitation on lying down with frequent belching. Burning and pressure between shoulder blades. Sleepy early in the morning but sleepless at night. Awakens at 3 A. M. Generally worse in morning; in dry weather; better in wet weather.

Opium.

Cases calling for this remedy will generally be characterized by drowsiness, constipation, contracted pupils. The breathing may be slow and snoring; the face dark red and hot; urine suppressed.

SPECIAL INDICATIONS :—

Stupor, unconscious. Imagines parts of the body very large. Thinks she is not at home. Dull, stupid as if drunk. Pupils contracted. Feels as if eyes were too big. Hearing acute, hears everything at night. Face dark red, bloated; purplish. Vomiting first of food then of fecal smelling substance, hiccough, cold face; complete stoppage of bowels. Squeezing pains in bowels, as if something was forced through a narrow place. Involuntary stools after fright. Cholera infantum with stupor, snoring, convulsions. Constipation, stools round, hard, black balls. No urine, bladder full. Short inspiration; long, slow expiration. Deep, snoring, breathing with wide open mouth. Numbness and insensibility. Heavy, stupid sleep. Bed feels so hot can hardly lie on it. Suitable in most cases of bad effects from fright.

Phosphorus.

SPECIAL INDICATIONS:—

Great indifference; answers no questions. Restlessness during a thunder-storm. Sensation of coldness in side and back of head. Beating in left temple. Headache over left eye; every other day. Top of head hot, after grief. Scaly, bald spots on head. Black spots pass before the eyes. Momentary blindness. Letters look red when reading. Green halo around the light. Small burning spots on eye-ball. Polypus of the nose; bleeds easily. Freckles on nose. Circumscribed red spots on face. Eyes sunken with dark rings around them. Cankerous sores on roof of mouth. Mouth sore, bleeds easily. Saliva increased, salt or sweetish. Hunger, must eat during chill. Spits up food without nausea. Vomits sour, offensive fluid looking like water, ink and coffee grounds. As soon as water becomes warm it is thrown up. Food comes up scarcely swallowed. Gone feeling in abdomen. Coldness in abdomen. Diarrhea. Constipation. Urine like curdled milk with brick-dust sediment and film on the top. Amenorrhea with spitting of blood, bleeding from anus or bladder. Menses suppressed with milk in breasts. Cough. Inflammation of the lungs. Yellow spots on chest. Great pressure in middle of breast-bone. Burning in small of back, in a small spot. Palms of hands burn. Epilepsy with consciousness. Sleepy all day, restless all night; especially before midnight. Hemorrhage from internal organs. Small wounds bleed much. Brownish spots. Tall, slender persons, easily magnetized; desire to be magnetized. Symptoms worse before and during thunderstorm.

Calcarea carb.

A valuable remedy for the ailments of scrofulous persons. Well adapted to children who sweat about the head, who have large stomachs and abdomens, whose bones are small, and who are troubled with swelling of the glands of the neck, (*kernels*, they are sometimes called); also large, fair, plump women or children, with light complexions, who tire out easily.

SPECIAL INDICATIONS—Delirium tremens, with talk about fire, rats, mice and murder. Fear of losing one's reason, or that people will notice one's confusion of mind. Dizzy when walking in the open air, or going up stairs. One-sided headache with much empty eructation. Headache begins in back of head and spreads to top of head, so severe she thinks she will go crazy. Icy coldness of the head; on one side. Scratches the head impatiently on being aroused from sleep. Dimness of vision, one side of an object obscured. Quivering of upper eye-lid. Cracking in ears when chewing. Swelling in front of left ear, painful. Nose swells at root. Polypus of nose. Upper lip swells in morning. Glands under tongue swell. Canker sores. Great desire for eggs; for wine; for salt or sweet things. Hungry in morning. Milk disagrees. Sour belchings; heart-burn. Pressing in the stomach as from a load or stone, after a moderate meal. Can't bear anything tight about the waist. Cold feeling in the abdomen. Hard lumps or kernels in the abdomen of children. Hard swellings in the groins. Stools grey or clay-like; hard and large. Menorrhagia. Before menses, swelling and painfulness of the breasts. Leucorrhea. Cough from feeling of plug or feather in the throat. Expectoration saltish or sweetish. White swelling of leg and foot with sensation of coldness. Cramp in hollow of knee when stretching out the leg. Cramp in the soles. Feet cold and damp. Foot-sweat makes feet sore. Elevated, red stripes on leg, with itching and burning after rubbing.

––––––

Dulcamara.

This remedy will be found useful in many troubles arising from taking cold by being in a cool damp place, as a cellar, vault, &c.; or from cold, damp weather. It suits best phlegmatic, torpid or scrofulous constitutions.

SPECIAL INDICATIONS :—

Sensation of chilliness over the back of head and over the back; feels as if the hair stood on end. Ring-worms on the scalp; glands about the throat swollen. Thick crusts on the scalp causing the hair to fall out. Scrofulous sore eyes every time one takes cold. Takes cold from every exposure. Thick, brownish-yellow crusts on face, forehead and chin; milk-crust. Lips twitch when in the cold air. Toothache from cold, especially with diarrhea. Itching and crawling on tip of tongue. Tongue and jaws become lame if one becomes chilled. Sore throat from every cold change. Colic, nausea, griping and diarrhea from taking cold. Urinary troubles from taking cold. Menses suppressed by cold. Rash on the body before the menses. Eruption on breasts. Mothers have eruptions after weaning. Asthma worse from cold, damp weather. Cough with looseness of the bowels. Stiff neck, pain in back and loins after taking cold. Coldness in small of back. Rheumatism after getting wet, severe pain when remaining in one position; must move about to get relief. Eruption oozes a watery fluid, bleeds after scratching; ceases to itch after scabbing over; worse from washing. Hives, worse in warmth, better in cold; after scratching it burns. Warts on back of hands.

Ignatia.

This remedy is especially suitable to nervous, hysterical females, of mild dispositions, who are easily excited either to mirth or sadness. Ailments from grief, mortification and disappointed love. It is also useful in the disorders of nervous children; for children who awaken in the night frightened, or who are afraid to sleep alone in the dark; for the spasms of children after being scolded, punished or frightened; for females who sigh much and are sad.

SPECIAL INDICATIONS:—

Changeable disposition; now joking and laughing, then sad and shedding tears. Pressing headache over the root of the nose, must bend the head forward for relief. Headache as if something hard pressed upon the surface of the brain. Throbbing pain in back of head, worse when straining at stool. Pain as if a nail were driven out through the side of the head, better when lying on that side. Headache from coffee, tea, tobacco, ardent spirits or nervous excitement, from sunlight, from noise. Cannot bear noise. Face alternately red and pale. Redness and heat of one cheek and ear.

Bites the cheek or tongue when talking or chewing. Stitches in the throat between the acts of swallowing, not during. Sensation of a lump in the throat when not swallowing. Throat feels worse when not swallowing; can swallow solid food better than liquids. Hunger and nausea at the same time. Weak, gone feeling at pit of stomach. Contractive pain in rectum one or two hours after stool. Stitches from the anus up the rectum. Prolapsus of the rectum from moderate straining at stool. Profuse discharge of watery urine. Cough after warm drinks; when standing still during a walk. Palpitation at night and morning in the bed. During the sleep of children, chewing motion of the mouth, jerks and grinding of the teeth.

Kali carbonicum.

SPECIAL INDICATIONS :—

Absent minded. Weeps much. Fear of being alone, fears she will die. Stitching pains in the head; in the eyes and root of the nose. One-sided headache, with nausea. Headache from riding in cars or carriage. Painful tumors on scalp, like blood boils. While using the eyes, sharp stitches in them. Sparks and spots before the eyes. Upper lids swollen. Headache and noise in ear after cold drink. Stitches in ear from within outward. Right ear hot, left, pale and cold. Cold in the head, with headache and backache. Nose open out doors, stopped in the room. Offensive yellow-green discharge from one ear. Nose bleed when washing the face ;

every morning at 9 o'clock. Teeth ache only when eating. Pain in back when swallowing. Food goes down only half-way, with gagging and vomiting. Desire for sugar. Stitches in pit of stomach. Stitching, shooting pains in abdomen. Constipation, stool large and dry ; feels distressed or sometime before stool. Menses too early and scanty ; cause an eruption on the thighs. Before and during menses, nettle-rash, great pain in back and down the back of the thighs. Asthma, worse in morning, must lean forward. Cough worse at 3 or 4 A. M. Stitches in the chest. Backache. Purple spots on arms and hands. Insides of hands and soles of feet itch. Yellow, scaly spots on abdomen ; around the nipples. Inclined to be fat.

——

Pulsatilla.

SPECIAL INDICATIONS :—

Easily moved to tears or laughter. Sadness in the morning, full of domestic cares. Mild, gentle, yielding, tearful dispositions. Headache, with a sort of humming in the head, and chilliness ; worse when quiet. Head aches worse in a warm room ; better when moving in the open air. Like a veil before the eyes ; better rubbing or wiping them. Styes especially on upper lid. Deafness; from cold after having hair cut. Can hear better on the cars or where machinery is running. Earache, Loss of smell. Green, fetid ; or thick yellow discharge from nose. Lower lip swollen and cracked in middle. Toothache worse from warmth, better when moving about in open air. Loss of taste.

Foul taste in morning. Clammy taste. Bitter taste after swallowing food or drink. Tongue feels as if burned in middle, even when moist. Tongue parched, dry, yet no thirst. Constant spitting of frothy, cotton-like mucus. Nausea or vomiting caused by fruits, fat things, pastry, ice-cold things. Aversion to fat food, pork, meat, bread, milk. Weight in stomach in morning. Pressure after every meal. Beating in pit of stomach. Pressure in abdomen and small of back as from a stone. Diarrhea usually at night. Constipation after taking quinine. Stools flat in shape, small size. Cannot retain the urine, it escapes sitting or walking, when coughing, passing wind, during sleep, the latter especially with little girls. Dropping of blood at end of urination. Orchitis, Prolapsus uteri. Menses suppressed. Leucorrhea. Lumps in breasts of girls before puberty. Breasts swollen, with rheumatic pains which change from place to place. Loss of voice at every excitement. Cough, dry at night, goes off when sitting up in bed; loose during day. Beating through the chest prevents sleep. Burning about the heart. Palpitation from excitement. Pains shift from place to place, worse at night, from warmth, better from uncovering. Hysterical; symptoms constantly changing. Fainting fits; paleness; shivering. Heat, chill, or sweat one-sided. Pains appear suddenly, leave gradually. Beating through the whole body. Nettle rash. Persons with light hair, blue eyes, pale face, chilliness. Women and children.

Rhus tox.

SPECIAL INDICATIONS:—

Absence of mind. Answers correctly but slowly. Thoughts of suicide; wants to drown himself. Dizzy. Headache. Erysipelas. Earache. Fever-blisters under the nose. Tip of nose red and sore. Nose puffy. Milk-crust. Stiffness of jaws; they crack when chewing. Fever-blisters around the mouth. Stinging at root of nose, extending to bones of face. Bread tastes bitter. Coppery taste. Tongue dry, red, cracked; triangular red tip; white on one side. Sore throat, feels stiff and lame. Nausea with great appetite. Stinging, beating in pit of stomach. Colic; must walk bent; from getting wet; worse at night. Diarrhea, with tearing pain down the thighs. Hoarse from straining the voice. Dry, hard cough, worse till midnight; back and limbs stiff. Stitches in chest worse when quiet. Pains about the heart with numbness and lameness of left arm. Palpitation when quiet. Stiff neck. Backache. Rheumatic pains, worse in cold, wet weather, in bed, and when quiet. Pains with numbness and crawling sensation. Pains worse on beginning to move; better from continued motion, Yawning, with pain in jaw. During fever, nettle rash.

Sanguinaria.

SPECIAL INDICATIONS :—

Irritable; morose. Dizziness when quickly turning and looking up; when lying down at night. Headache begins in back of head and settles over right eye. Head sore in spots. Veins of temples enlarged; feel sore when touched. Eyes feel as if hairs were in them. Profuse lacrymation with cold in the head, sore eyes. Burning of the ears, red cheeks. Polypus of nose. Catarrh, running of water from nose, sneezing; worse on right-side. Hay-fever. Sick and faint at smell of flowers. Veins of face distended, face feels stiff. Circumscribed red cheeks. Neuralgia of face. Lips swell toward evening. Upper lip swollen, burns, hard and blistered. Jaws stiff. Toothache when picking the teeth, or in a hollow tooth when touched by the food. Loss of taste, tongue feels burnt. Sweet things taste bitter. Tip of tongue burns as if scalded. Throat burns after eating sweet things. Sore throat. Heat in throat. Diphtheria; pearly coating on palate. After eating, soon feels empty; water brash. Intense nausea in paroxysms, worse when stooping; followed by nettle rash, heart-burn, headache, chills. Burning in stomach with headache. Jerking in stomach as if from something alive. Heat streaming from chest to liver. Jaundice. Feeling as if hot water were poured from chest to abdomen, afterward diarrhea. Diarrhea after cold in the head. Flashes of heat and leucorrhea at change of life. Loss of voice, with dryness, soreness, swelling and redness. Urging, without result; passes flatus only; feels as if there was a mass in the rectum. Chest diseases, with a red spot on one or both cheeks. Whooping cough, worse at night with diarrhea. Burning in breastbone. Soreness under right breast. Rheumatism. Burning in palms of hands, and soles of feet, worse in bed. Prickling sensation of warmth spreading over body. Veins distended, and feel sore. Nettle rash, afterward nausea. Eruptions on face of young women with deficient monthly flow.

Sepia.

This remedy seems to have an especial effect upon the female organism. It is very valuable for many of the ailments of women during pregnancy and nursing. It is one of the best remedies in *nervous* headaches, particularly when occuring at about the time of the menses.

SPECIAL INDICATIONS:—

Sad about one's health or about her domestic affairs. Dreads to be alone. Great indifference to one's own family, to those one loves the best. One-sided headache, (left side mostly;) better in open air and when lying on painful side. Beating headache in back part of head, worse from motion, turning the eyes or lying on back. Involuntary jerking of head backward and forward. Sensation of coldness on top of head. Dullness of sight; sees but one-half of an object clearly. Green circle around the candle-light. Upper lids seem heavy. Eruption on the lobe of the ear; on the nape of the neck; on the tip of nose. Small red pimples on forehead. Yellowness of face, around the mouth and across the nose. Swelling of upper lip. Toothache. Food tastes too salt. Sore throat, worse left side. Sensation of plug or lump in throat. Aversion to meat. Desire for vinegar; for wine. Belchings which taste like rotten eggs. Nausea from smell of food, or when riding in a carriage. Empty feeling in stomach and abdomen. Beating at pit of stomach.

Brown spots on abdomen. Oozing of moisture from rectum. Sore between buttocks. Sensation of lump or weight in rectum. Blood-red urine with a scum on the surface and a white sediment. Sediment adheres to sides of vessel, can hardly be removed. Cough in bed till midnight. Palpitation, after emotions of the mind, after dinner; wakes up at night with violent beating. Brown spots on chest. Arms and fingers fall asleep. Skin on palms peels off. Lower limbs icy cold. Crackling in knee joint. Feet sweat, causing an offensive odor and soreness between the toes. Knees weak. Cramps in buttocks, at night in bed. Limbs twitch during sleep. Awakes from sleep at 3 A. M.; cannot sleep again. Moist eruption in bend of knees. Ringworms. Especially suitable for females with dark hair and mild dispositions.

––––

Podophyllum.

No remedy will be more frequently used for loose discharges from the bowels. It is one of the first to be studied in *cholera morbus, cholera infantum* and ordinary diarrhea. The stools are often painless, gushing, bilious, are apt to be most profuse in the morning, frequently driving the patient out of bed quite early.

SPECIAL INDICATIONS:—

Dizzy when standing in the open air; with a sensation of fullness over the eyes, and a tendency to fall forward. Child's head is hot; rolls head from side to side, moaning, with diarrhea while cutting teeth. Headache alternates with diarrhea

Headache in morning with flushed -face. Child clenches the teeth together, grinds the teeth at night. Sore throat, worse on left side, in the morning and when swallowing liquids. Belchings taste like rotten eggs. Spitting up food. Infants vomit up milk ;˙ they have protrusion of the anus. Pain, soreness and fulness in the region of the liver ; patient desires to rub and stroke the abdomen with the hand. Sharp pain in right groin during pregnancy. Diarrhea while being washed. Prolapsus-ani from slight exertion. Passes urine frequently during the night; passes urine during sleep. Prolapsus-uteri. Leucorrhea. Menses suppressed with pain in lower part of back. Cough during *intermittent fever.* Pain between shoulders in morning. Flashes of heat up the back. Backache in females ; with the piles. Backache before chill. Delirious talking during fever. Jaundice.

Phytolacca.

This remedy I have used but little except in diseases of the throat and breast, and in rheumatism. It is deemed a specific for "Caked breast." A bad case of diphtheria recovered rapidly under the use of *Phytolacca* and *Kali-permanganicum.* I give it in rheumatism when I cannot decide what remedy is indicated, and often with the best result.

SPECIAL INDICATIONS :—

Pain shoots from left eye to top of head. Sick headache comes every week ; with backache and bearing down. Sharp pain goes through the eyeball when reading or writing. Motion of one eye independent of the other. Sensation of soreness deep in the brain. Shooting pains through both ears when swallowing, worse on right side. Cold in the head; stoppage of one nostril, both stopped when riding. Blotches on the face, worse after eating or washing the face. Left ear and left side of face swollen like erysipelas; spreads over the scalp. Convulsions, head drawn forward so the chin touches the breast-bone. Tongue fiery-red at tip and smarting; coated thick at back part. Mouth full of tough, stringy saliva. Ulcers on inside of right cheek. Sore throat, with severe pains in head, neck and back; faint on rising up. Diphtheria. Hungry soon after eating. Vomiting of clotted blood and slime, with great pain and desire for death to relieve the suffering. Diarrhea early in the morning. Stools of mucus and blood; look like the scrapings of intestines. Urinary disorders. Urine leaves a red sediment and stain on the vessel. Menses too frequent and profuse, with pain in the breasts. Breasts and nipples inflamed and sore. When the child nurses pains seem to start from nipple and extend all over the body. "Gathered Breast." Breast full of hard lumps. Pain about the heart, extends into right arm. Pains streaking up and down the spine. Rheumatism; pains worse down outside of thigh. Pains in middle of long bones; shift about. Worse in damp weather and at night. Boils, when coming near ulcers.

––––

Cuprum.

This remedy will be found useful in *asthma*,

whooping cough, *colic*, and in spasmodic affections of children generally. Also in *cholera* and other ailments when characterized by *cramps* in the muscles.

SPECIAL INDICATIONS:—

Strange crawling or tingling sensation on top of the head. Brain diseases in children, with convulsive movements of the extremities. Eye-balls red, move from side to side. Nose feels as if there was a great deal of blood in it; bleeds on right side. Sweetish taste in mouth. Vomiting relieved by a drink of cold water. Hiccough precedes vomiting. Asthma. Nausea, vomiting and cramps during menses, or with menses suppressed. Cramps in abdomen; abdomen drawn in. Constipation; nothing is able to pass; hiccough, colic; vomits matter that ought to pass the other way. Asthma before menses. Cough, better from drinking cold water. Whooping cough. Eruptions in bend of elbow, with yellow scabs; itching worse at night. Cramps in limbs. Convulsions; limbs are stiff, extended, spread apart; mouth opened. Convulsions begin in a finger or toe. Great itching without eruption.

——— .

Gelseminum.

SPECIAL INDICATIONS:—

Dull, quiet, desires to be let alone. Complaints from bad or exciting news. Conscious, but cannot move, pupils dilated, eyes closed. Dizziness with dilated pupils and dim sight. Child dizzy when carried, seizes hold of the nurse for fear of falling. Headache; severe pain in forehead and top of head, dim sight, roaring in ears. Head feels enlarged. Headache alternates with pains in lower part of abdomen. Headache, with great hunger before and during. Brain feels as if sore. Dull heavy pain in back of head; drowsy; eyes heavy and red, half-closed. Head feels heavy. Headache begins in nape of neck, extends over head, to forehead and eye-balls, with a bursting feeling. Eyelids heavy. Double vision. Sudden transient loss of hearing. Troubles of teething; child's face deep red. Tongue trembles. Diarrhea after sudden emotion or any excitement. Copious flow of urine relieves the headache. Cannot retain the urine; nervous children. Long crowing inspiration; sudden, forcible expiration. Fears that her heart will cease beating, unless she moves constantly. Chill, yet the surface is warm. Neuralgia, shooting pains along the nerve; worse from any change in the weather. Stupor. Fevers, with a tendency to convulsions.

Sulphur.

A remedy especially useful in chronic ailments, eruptions, constipation, piles, disorders of digestion. It is often called for in obscure cases, particularly where there is, at times, a tendency to eruptions; or where such tendency may have existed in the parents of patient.

SPECIAL INDICATIONS:—

Weak memory for names. Foolish delirium; thinks every thing is beautiful, even rags and filth. Disgusting smell from the body. Dizzy, inclined to fall to left. Tearing, or stitches in forehead, from within outward. Tight feeling in the head when doing mental work. Sensation of emptiness in back part of head. Throbbing headache at night. Soreness and heat on top of head. Eruption on back of head and behind ears. Dandruff, head itches worse at night when warm in bed. Aversion to the light, with stitches in eyes, worse in sultry weather. Sore eyes, worse from bathing the eyes. Humming, hissing, swashing as from water in the ears. Red ears. Nose runs, outdoors; stopped; in-doors. Freckles and black pores on nose. Eruption around chin and corners of mouth. Upper lip swollen. Toothache, left side, from open air, washing with cold water; with stitches in the ears. Profuse saliva. Sensation of lump in throat. Sensation of a hair in throat. Great craving for food, with little children. Fulness from eating little. Weak, faint feeling at pit of stomach about 10 or 11 A. M. Painful swelling of glands in the groins. Children with large bellies and small legs. Diarrhea, worse in morning. Soreness of anus. Looks red around the anus. Stools make the parts raw. Constipation, stools hard, knotty and insufficient. Piles, blind or bleeding. Stinging, burning and itching around the anus. Before menses: cough; headache; nose bleed. Menses too scanty; too late; too short duration, or too early and profuse. Nipples smart and bleed after nursing. Abscess of the breast. Asthma, worse lying on feathers. Raises green lumps of sweet taste. Sensation of coldness in the chest. Cracking in neck. Hang-nails. Chilblains. Cramps in calves; in soles. Feet burn, wants them uncovered. Corns ache and sting. Pains in limbs, worse lying on a feather bed. Easily awakened; short naps. Sweats on nape of neck. Children grow poor, look very old and wrinkled. All sorts of eruptions. Ail-

ments are generally worse from washing in cold water and from warmth of bed.

———

China.

This remedy is indicated for the prostration following the loss of bodily fluids, as profuse menses, long continued nursing, excessive discharge of pus, leucorrhea &c.

SPECIAL INDICATIONS:—

Delirium after loss of blood; sees persons when closing the eyes. Fainting, loss of sight, ringing in ears, after hemorrhage. Stitches from temple to temple. Headache, worse from slightest touch, from open air, from draught of air; better from hard pressure. Ringing in the ears. Nose bleed in the morning. Veins on forehead and hands distended. Toothache while nursing an infant. Bitter taste in back part of throat. Hunger at night. Appetite less in foggy weather. Fulness in stomach, belching does not relieve. Diarrhea, stool painless, frothy, with fermentation in bowels; stool dark colored, black. Profuse menses. Pressure like a stone between the shoulder blades. Pains in limbs and joints, worse from slight touch. Thirst before and after chill, not during; increased thirst during sweat. Worse every other day.

INDEX AND VOCABULARY.

————◆◆————

In the following Index and Vocabulary will be found in alphabetical order the subjects treated of, together with such words as may require explanation.

www.ingramcontent.com/pod-product-compliance
Lightning Source LLC
Chambersburg PA
CBHW021809190326

41518CB00007B/514